SpringerBriefs in Computer Science

Series Editors
Stan Zdonik
Peng Ning
Shashi Shekhar
Jonathan Katz
Xindong Wu
Lakhmi C. Jain
David Padua
Xuemin Shen
Borko Furht
VS Subrahmanian

For further volumes:
http://www.springer.com/series/10028

Dawei Li • Jie Wu

Energy-aware Scheduling on Multiprocessor Platforms

 Springer

Dawei Li
Department of Computer
  and Information Sciences
Temple University
Philadelphia, PA, USA

Jie Wu
Department of Computer
  and Information Sciences
Temple University
Philadelphia, PA, USA

ISSN 2191-5768   ISSN 2191-5776 (electronic)
ISBN 978-1-4614-5223-2   ISBN 978-1-4614-5224-9 (eBook)
DOI 10.1007/978-1-4614-5224-9
Springer New York Heidelberg Dordrecht London

Library of Congress Control Number: 2012950310

Printed on acid-free paper

Springer is part of Springer Science+Business Media (www.springer.com)

Preface

Multiprocessor platforms play an important role in modern computational systems, and appear in various applications, ranging from energy-limited hand-held/battery-powered devices to large data centers. As their performance increases, energy consumption in these systems also increases significantly. Dynamic Voltage and Frequency Scaling (DVFS), which allows processors to dynamically adjust the supply voltage and the clock frequency to operate on different power/energy levels, is considered an effective way to achieve the goal of saving energy. Recently, energy-aware task scheduling on DVFS multiprocessor platforms has been a hot topic. Our work in this book surveys existing researches that have been done on this topic. We notice that energy-aware scheduling problems are intrinsically optimization problems, the formulations of which greatly depend on the platform and task models under consideration. Thus, we classify existing works according to two key dimensions, namely, homogeneity/heterogeneity of multiprocessor platforms and the task types under consideration. Under this classification, other sub-issues are also included in this book, namely, slack reclamation, fixed/dynamic priority scheduling, partition-based/global scheduling, task preemption/non-preemption and application-specific power consumption, etc. Our work provides an overall and comprehensive survey on energy-aware scheduling on multiprocessor platforms.

Philadelphia, PA, USA   Dawei Li and Jie Wu

Contents

Chapter 1
Introduction

Abstract This chapter introduces the energy consumption issue in modern computational multiprocessor platforms and describes the basic concept of Dynamic Voltage and Frequency Scaling (DVFS). For ease of understanding and further discussions on multiprocessor platforms, important concepts related to energy-aware scheduling on uniprocessor platforms are also made clear first. For multiprocessor platforms, we mention that several new issues will also be considered besides the ones considered for uniprocessor platforms. The organization of the book is provided in the last paragraph.

The primary design goal of computational systems has been about system performance improvement, where performance is often characterized by processing speed for a given task. However, as the performance increases, energy consumption in these systems also increases significantly. High energy consumption becomes a key problem; in the mobile energy-limited devices, it results in a short lifetime, and in large data centers, it results in high electricity bills. Thus, energy-aware task scheduling is drawing more and more attention.

Dynamic Voltage and Frequency Scaling (DVFS), which allows processors to dynamically adjust the supply voltage or the clock frequency to operate on different power/energy levels, is considered an effective means of achieving the goal of saving energy. DVFS is also called Dynamic Voltage Scaling (DVS) in various other literature, because, in most cases, the supply voltage has a one-to-one corresponding relation with the operating frequency, and thus dynamically adjusting the supply voltage is equivalent to dynamically adjusting the clock frequency. Throughout this book, for consistency, and to eliminate any confusion, only the term DVFS is used.

In past decades, energy-aware task scheduling in DVFS uniprocessor systems attracted a lot of research interest. Generally, the active power consumption, when a processor is running tasks, includes both dynamic power (due to switching activity) and static power (due to leakage current). When static power consumption is negligible, the active power consumption of a processor can be approximated

D. Li and J. Wu, *Energy-aware Scheduling on Multiprocessor Platforms*,
SpringerBriefs in Computer Science, DOI 10.1007/978-1-4614-5224-9_1,
© The Author(s) 2013

as proportional to the cube of the clock frequency, while the actual execution time of a CPU cycle is just inversely proportional to the clock frequency. Thus, the energy consumption per CPU cycle is proportional to the square of the clock frequency. For a given task, its computation requirement is quantified as the required number of CPU cycles and is regarded as fixed, more or less. Hence, slowing down the processing speed (clock frequency) as much as possible, without missing its predefined deadline, will significantly reduce the total energy consumption. When static power is non-negligible, using a higher frequency to execute tasks and then put the processor in shutdown/sleep mode, where power consumption is much less, may save more energy. Hence, a tradeoff between *slowdown* and *shutdown* is required.

Multiprocessor platforms are playing important roles in modern computing systems, and they significantly improve the performance of computational systems. Consequently, energy consumption in these systems also increases significantly. Recently, energy-aware scheduling in multiprocessor DVFS systems has become a hot research topic in both academic and industrial societies. In addition to slowdown and shutdown, in multiprocessor systems, it is also important to determine the optimal number of processors that should be used. This is because, on the one hand, by using more processors, these processors can operate at lower frequencies, leading to the reduction of dynamic energy consumption; on the other hand, using more processors will increase the static energy consumption. Besides, determining how to assign tasks to different processors, and deciding whether and to which processor to migrate tasks (if migration is allowed), are all important issues in energy-aware scheduling on multiprocessor platforms. Such issues in energy-aware uniprocessor scheduling as task preemption, slack reclamation, and priority constraints, etc., also become more complicated on multiprocessor platforms.

Generally, energy-aware scheduling problems on multiprocessor platforms can be formulated as optimization problems, the formulations and solutions of which greatly depend on the platform and task models that are under consideration. Thus, the classification of existing works discussed in this book are mainly based on the platform types and task models considered.

In this book, we will first present the system models dealt with in most research works in Chap. 2, which consists of task models, platform models, and other concepts related to energy-aware scheduling on multiprocessor platforms. A comprehensive survey of energy-aware scheduling problems are presented for homogeneous platforms and heterogeneous platforms in Chaps. 3 and 4, respectively. For homogeneous platforms in Chap. 3, problems are further classified according to four task models, namely, frame-based tasks, tasks with precedence constraints, periodic tasks, and sporadic tasks. For heterogeneous platforms, since, to the best of our knowledge, little has been done for energy-aware scheduling of sporadic tasks, we only survey problems for frame-based tasks, tasks with precedence constraints, and periodic tasks. Some works that are similar to ours are provided and discussed in Chap. 5. The conclusion and our future directions are presented in Chap. 6.

Chapter 2
System Model

Abstract It is obvious that energy-aware scheduling problems largely depend on the tasks and platform under consideration. This chapter provides the system models that we consider in this book. Task models are presented in Sect. 2.1, which includes four types of tasks, namely, frame-based tasks, tasks with precedence constraints, periodic tasks, and sporadic tasks. Uniprocessor power consumption models are provided in Sect. 2.2. Multiprocessor platform models are presented in Sect. 2.3. Section 2.4 discusses other concepts and assumptions related to energy-aware scheduling on multiprocessor platforms. These concepts and assumptions are also important for energy-aware scheduling problems.

2.1 Task Models

Actually, various types of tasks/applications are considered for energy-aware scheduling. In this book, four typical task models are included; they are *frame-based tasks*, *tasks with precedence constraints*, *periodic tasks*, and *sporadic tasks*.

Frame-based Tasks: Frame-based tasks are a set of tasks, $\mathcal{T} = \{\tau_1, \tau_2, \cdots, \tau_n\}$ that are released at the same time 0 and share a common deadline D. The execution requirement of task τ_i is denoted by C_i, which is defined as the Worst Case Execution Time (WCET) at the processor's maximum frequency. Denote $WCEC_i$ as the Worst Case Execution Cycles (WCECs) of task τ_i; then, $C_i = WCEC_i / f^{max}$. For tasks with precedence constraints, C_i is defined with the same meaning. The utilization of task τ_i is defined as $u_i = C_i / D$.

Tasks with Precedence Constraints: Tasks in the set $\mathcal{T} = \{\tau_1, \tau_2, \cdots, \tau_n\}$ that have precedence constraints are modeled by a weighted Directed Acyclic Graph (DAG). Let the DAG be (V, E); the node set V of this graph corresponds to tasks $\tau_1, \tau_2, \cdots, \tau_n$. The edge set E corresponds to precedence constraints. The weight of each node, C_i represents the execution requirements of task τ_i. The overall

D. Li and J. Wu, *Energy-aware Scheduling on Multiprocessor Platforms*,
SpringerBriefs in Computer Science, DOI 10.1007/978-1-4614-5224-9_2,
© The Author(s) 2013

Fig. 2.1 An example of tasks
with precedence constraints

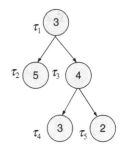

Fig. 2.2 A valid schedule on
two processors

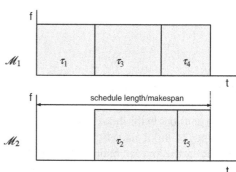

completion time of scheduling these tasks on multiple processors is called the
makespan or *schedule length*, denoted by D. The utilization of task τ_i is defined
as $u_i = C_i/D$.

Consider an example of five tasks with worst case execution times, $C_1 = 3$,
$C_2 = 5$, $C_3 = 4$, $C_4 = 3$, and $C_5 = 2$. Their precedence constraints are shown in
Fig. 2.1. A valid schedule with schedule length of 10 is shown in Fig. 2.2.

With the energy-aware consideration, two problems are usually associated with
scheduling framed-based tasks and tasks with precedence constraints: minimizing
the schedule length with an energy consumption constraint and minimizing energy
consumption with a schedule length constraint.

Periodic Tasks: A periodic task is an infinite sequence of task instances (or called
jobs), where each job/instance of a task comes in a regular period. Each task τ_i in a
periodic task set $\mathscr{T} = \{\tau_1, \tau_2, \cdots, \tau_n\}$ is described by (A_i, C_i, D_i, T_i), where, A_i is its
initial arrival time, C_i represents the worst cast execution time of its one job/instance,
D_i represents its relative deadline, and T_i represents its period. If, initially, all of the
tasks are released at the same time 0, which means $A_i = 0, \forall i, 1 \leq i \leq n$, then A_i
is omitted; task τ_i is simply denoted by (C_i, D_i, T_i). The jth job of τ_i is denoted by
$\tau_{i,j}$. Besides, some useful terms are defined for periodic tasks. The utilization of a
periodic task is defined as $u_i = C_i/T_i$. The total utilization of a task set is denoted
by $U^{total} = \sum_{\tau_i \in \mathscr{T}} u_i$. The tasks are considered to have *implicit* deadlines if $D_i = T_i$
and are considered to have *constrained* deadlines if $D_i \leq T_i$. Oftentimes, the hyper-
period H of a task set is defined as the minimal common multiplier of all periods

Fig. 2.3 Examples of a periodic task and a sporadic task. (**a**) An example of a periodic task. (**b**) An example of a sporadic task

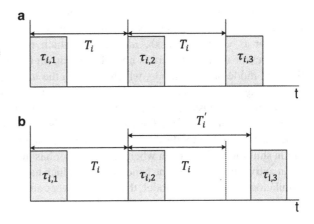

of the tasks. Then, the number of task τ_i's jobs during the hyper-period is H/T_i. Besides, a task τ_i's density is defined as $\lambda_i = C_i/D_i$. A typical example of a periodic task τ_i with period T_i is shown in Fig. 2.3a. In the research community, a set of periodic tasks are generally considered as a whole.

Sporadic Tasks: A sporadic task is also an infinite sequence of task instances/jobs, where job arrivals have a minimal inter-arrival time rather than a fixed period. Obviously, a periodic task is a special kind of a sporadic task. Because of this relation between periodic tasks and sporadic tasks, the representation of periodic tasks can also be used for sporadic tasks, with the slight difference that T_i represents the sporadic task's minimal inter-arrival time. Some research approaches and results for periodic tasks can also be used for sporadic tasks. A typical example of a sporadic task τ_i with minimal inter-arrival time T_i is shown in Fig. 2.3b. Notice that T_i is just the minimal inter-arrival time. Practical inter-arrival times can be greater than T_i. As shown in Fig. 2.3b, the inter-arrival time between $\tau_{i,2}$ and $\tau_{i,3}$ is $T_i' > T_i$. Generally, a set of sporadic tasks are considered as a whole.

2.2 DVFS Models

Our work in this book focuses on energy-aware scheduling on multiprocessor platforms. In some sense, multiprocessor platforms can be regarded as consisting of multiple uniprocessors. In this section, we will present the power consumption model for a single processor first. Both homogeneous and heterogeneous multiprocessor platform models will be described in the next sections.

The practical power consumption model of a DVFS processor is quite complex and varies among different manufacturing technologies.

Generally, a DVFS processor can be in three kinds of operation modes: *active* mode, also called busy mode or run mode, which refers to the mode in which the processor is executing tasks; *idle* mode, which refers to the mode in which the processor is on, but there is no task running on it; *shutdown* mode, also called dormant mode or sleep mode, which refers to the mode in which the processor is not running any tasks, is put in a very low power state, and can be recalled to active mode when new tasks arrive. If a processor can be put into a shutdown mode, it is called a *shutdown-enabled* processor; otherwise, it is called a *shutdown-disabled* processor. A shutdown-disabled processor has to stay in idle mode and cannot be put in shutdown mode, even when the processor has no task to execute.

In **active** mode, we use the general model which has been verified with the SPICE simulation in [1]. In this model, the active power consumption is given by:

$$P^{act} = P^{dyn} + P^{sta} + P^{on}$$

where P^{dyn} is the dynamic power consumption due to switching activity, P^{sta} is the static power consumption due to leakage current, and P^{on} is the intrinsic power consumption needed to keep the processor on and is assumed to be a constant.

The dynamic power, P^{dyn}, is given by:

$$P^{dyn} = C_e V_{dd}^2 f$$

where C_e is the average switched capacitance per cycle, V_{dd} is the supply voltage, and f is the clock frequency.

The static power, P^{sta}, is given by:

$$P^{sta} = V_{dd} I_{subn} + |V_{bs}| I_j$$

in which I_{subn} is the sub-threshold leakage current, given by:

$$I_{subn} = K_3 e^{K_4 V_{dd}} e^{K_5 V_{bs}}$$

where K_3, K_4, K_5 are constants. V_{bs} is the voltage applied between body and source, and I_j is the reverse bias junction current. Besides, there is a relation between operating frequency, supply voltage, and threshold voltage:

$$f = (V_{dd} - V_{th})^\gamma / L_d K_6$$

where γ and K_6 are constants, and L_d represents the logic depth. The threshold voltage is given by:

$$V_{th} = V_{th1} - K_1 V_{dd} - K_2 V_{bs}$$

where all V_{th1}, K_1, K_2 are platform-specific constants.

Assuming that the minimum and maximum operating frequencies are f^{min} and f^{max}, respectively, there are two types of processors: a processor is called *ideal* if it can operate at any frequency in the range $[f^{min}, f^{max}]$, and it is called *non-ideal* if it can only operate at a set of discrete frequencies in this range. If it can operate on K different frequencies, denote $\{f^{(1)}, f^{(2)}, \cdots, f^{(K)}\}$ as the available frequency set. Without loss of generality, assume that $f^{min} = f^{(1)} < f^{(2)} < \cdots < f^{(K)} = f^{max}$. f^{min} can be or cannot be 0, and f^{max} can be or cannot be $+\infty$, according to different assumptions. For a processor executing at frequency f, its speed s can be defined as $s = f/f^{max}$.

This general active power consumption model is quite practical and complicated, but in most research, a lot of assumptions are made under different contexts, and the model is thus simplified based on this general model, as we will present later for separate papers.

In **idle** mode, the power consumption is denoted as P^{idl}, which may be much less than the active power. It's exact value is determined by different assumptions or by measuring practical platforms.

In **shutdown** mode, the power consumption is denoted as P^{sh}, which is much less than P^{idl}. It is assumed to be negligible in a lot of research literature or just a small percent of P^{idl} in other ones.

Since the processor can transit between different modes, there might be some transition overhead during the transitions. Generally, it is assumed that the transition between an active mode and an idle mode can be done immediately and requires no additional energy overhead. However, the transition between idle mode and shutdown mode may require both time and energy overhead, denoted by t^{ov} and E^{ov}, respectively.

2.3 Platform Models

This section will describe multiprocessor models, which consist of a homogeneous multiprocessor platform and a heterogeneous multiprocessor platform.

Homogeneous Platforms: Roughly speaking, if all of the processors on a multiprocessor platform are identical, this platform is regarded as a *homogeneous* platform. For homogeneous platforms, if all of the processors must operate at the same supply voltage and clock frequency at any time, it is called a *dependent* platform, which is the common case in a chip multiprocessor (CMP) or, in other words, a multi-core chip. For a dependent platform, although all of the processors must operate at the same supply voltage and clock frequency, they may transit into idle mode or shutdown mode independently. If processors on a platform can operate on different supply voltages and clock frequencies, and can adjust their values independently, it is called an *independent* platform. In addition, a new type of homogeneous platform is considered in previous and current researches. This platform is called a partitioned multi-core, where all cores on the platform are partitioned into different islands/blocks. Cores from the same block/island are dependent, while cores from different blocks/islands are independent.

Heterogeneous Platforms: If any processor is not identical to another one on a multiprocessor platform, this platform is deemed a *heterogeneous* platform. A typical example of a heterogeneous platform is one containing both DVFS processors and non-DVFS Processing Units (PUs), such as FPGA. For non-DVFS Processing Units (PUs), if its power consumption is dependent on the workload assigned to it, it is called a *workload-dependent* non-DVFS PU. Otherwise, it is called a *workload-independent* non-DVFS PU.

Also, a heterogeneous platform may consist of DVFS processors with different frequency ranges or different power consumption functions. Consider a heterogeneous platform with m DVFS processors, $\mathcal{M}_1, \mathcal{M}_1, \cdots, \mathcal{M}_m$. Since processors may have different configurations and power consumptions, the previously presented power consumption model should indicate this aspect by simply adding a processor index. Namely, P_j^{act}, P_j^{dyn}, and P_j^{sta} represent the active power, dynamic power, and static power of processor $\mathcal{M}_j$, $j \in [1, \cdots, m]$. P_j^{on} represents the intrinsic power to keep the processor $\mathcal{M}_j$ on. The frequency range of processor $\mathcal{M}_j$ is represented by $[f_j^{min}, f_j^{max}]$.

Tasks' worst case execution requirements are also dependent on which processor the tasks are assigned to. Thus, $c_{i,j}$ is used to represent the Worst Case Execution Time (WCET) of task τ_i when it is assigned to processor $\mathcal{M}_j$ at the maximum frequency f_j^{max}. $c_{i,j} = WCEC_i/f_j^{max}$. Similarly, $u_{i,j} = c_{i,j}/T_i$ is defined as the utilization of τ_i when it is assigned to $\mathcal{M}_j$. The utilization of processor $\mathcal{M}_j$ is the sum of utilizations of all the tasks that are assigned to it, denoted by $U_j = \sum_{\tau_i \in \mathcal{T}_j} u_{i,j}$, where $\mathcal{T}_j$ is the task set assigned to processor $\mathcal{M}_j$.

For shutdown-enabled processors, the time overhead and energy overhead of putting the processor into shutdown mode and returning it to active mode are denoted by t_j^{ov} and E_j^{ov}, respectively. Generally, heterogeneous platforms are assumed to be independent.

Notations that are consistently used in this book are provided in Table 2.1.

2.4 Other Related Concepts

This section will describe some important concepts that are related to energy-aware scheduling. They reflect some characteristics of practical platforms and tasks. When they are taken into consideration, more constraints and requirements should be included in the energy-aware scheduling algorithms.

Slack Reclamation: A processor may stay idle for a short period of time, even when it is allocated to execute some tasks, because of a task's late arrival or advanced completion. These idle periods of time are referred to as slacks. Generally, there are two kinds of slacks. One of them results from the scheduling scheme

Table 2.1 Notations used in this book

Notation	Description
$\mathcal{T}$	Task set $\{\tau_1, \tau_2, \cdots, \tau_n\}$
τ_i	The ith task in task set $\mathcal{T}$
n	The number of tasks in $\mathcal{T}$
$\tau_{i,j}$	The jth job of a periodic/sporadic task τ_i
A_i	Initial arrival time of task τ_i
$WCEC_i$	Worst case execution cycles of task τ_i for frame-based tasks or precedence-constrained tasks or WCEC of a job of periodic/sporadic task τ_i
C_i	Worst case execution time of task τ_i for frame-based tasks or precedence constrained tasks or WCET of a job of periodic/sporadic task τ_i
D	The common deadline of a frame-based task set $\mathcal{T}$
D_i	Relative deadline of task τ_i
T_i	Period of periodic task τ_i; or minimal inter-arrival time of sporadic task τ_i
u_i	Utilization of τ_i
U^{total}	Total utilization of task set $\mathcal{T}$
λ_i	Density of task τ_i
H	Hyper-period of a periodic task set $\mathcal{T}$
$\mathcal{M}_j$	jth processor/core or jth processor type
$P^{act}(P_j^{act})$	Active power consumption (of $\mathcal{M}_j$)
$P^{dyn}(P_j^{dyn})$	Dynamic power consumption (of $\mathcal{M}_j$)
$P^{sta}(P_j^{sta})$	Static power consumption (of $\mathcal{M}_j$)
$P^{on}(P_j^{on})$	Intrinsic power to keep processor ($\mathcal{M}_j$) on
$P^{idl}(P_j^{idl})$	Power consumption (of $\mathcal{M}_j$) in idle mode
$P^{sh}(P_j^{sh})$	Power consumption (of $\mathcal{M}_j$) in shutdown mode
$f^{min}(f_j^{min})$	Minimum frequency of processor ($\mathcal{M}_j$)
$f^{max}(f_j^{max})$	Maximum frequency of processor ($\mathcal{M}_j$)
s	Processor speed defined as f/f^{max}
$f^{(k)}(f_j^k)$	The kth frequency level of processor ($\mathcal{M}_j$)
$c_{i,j}$	WCET of τ_i when it is assigned to processor $\mathcal{M}_j$
$u_{i,j}$	Utilization of task τ_i assigned to $\mathcal{M}_j$
$\mathcal{T}_j$	Task set assigned to $\mathcal{M}_j$
U_j	Utilization assigned to $\mathcal{M}_j$
$t^{ov}(t_j^{ov})$	Switching time overhead (of $\mathcal{M}_j$)
$E^{ov}(E_j^{ov})$	Switching energy overhead (of $\mathcal{M}_j$)
m	The number of processors

itself, while the other one results from the difference between a task's Actual Case Execution Time (ACET) and Worst Case Execution Time (WCET). In energy-aware scheduling, the slacks can be used to slow down or shut down the processor to achieve the goal of saving energy.

Task Preemption: If the current task under execution can be preempted by higher priority tasks, this task is considered to be *preemptive*; otherwise, it is considered to be *non-preemptive*. Schedulability conditions for preemptive scheduling and non-preemptive scheduling are quite different, resulting in the difference between energy-aware preemptive scheduling and non-preemptive scheduling.

Fixed/Dynamic Priority Scheduling: If priority values of tasks will not change during runtime, the scheduling is considered to be of *fixed priority*; otherwise, it is considered to be of *dynamic priority*. Rate Monatomic (RM) scheduling, which ranks the tasks according to tasks' arrival frequencies/rates, is a typical example of fixed priority scheduling. Earliest Deadline First (EDF), which dynamically assigns priorities according to the task job's deadlines during runtime, is a typical example of dynamic priority scheduling.

Partition-based/Global Scheduling: The most commonly used approach of scheduling tasks on multiprocessor platforms is *partition-based scheduling*, where each task is assigned statically to one processor. Partition-based scheduling allows schedulability to be verified by well-understood single-processor analysis techniques. Partition-based scheduling also requires less scheduling overhead on practical platforms. On the other hand, we have *global scheduling*, in which there is a single job queue from which jobs are dispatched to any available processor according to a global priority scheme. Global scheduling allows different instances/jobs of the task to be executed upon different processors. Each instance/job can start its execution on any processor and may migrate during runtime from one processor to another if it gets preempted by a higher priority job.

Task Migration: Task migration means that a task need not be entirely executed on one processor. Given that task migration is allowed, for frame-based tasks and tasks with precedence constraints, one task's former part and the rest can be executed on more than one processor sequentially; for periodic tasks, a task's single job's former part and the rest can be executed on more than one processor sequentially; besides, jobs released early and later jobs can be executed on different processors. If task migration is not allowed, the whole part of a task must be executed on one processor. It is obvious that whether task migration is allowed or not has a great influence on a scheduling strategy. Consequently, it will also influence energy-aware scheduling.

Single Task Concurrent Execution (Parallel Task Execution): Traditionally, it is assumed that one job instance of a task is unable to be executed on more than one processor simultaneously, even when task migration is allowed. Unlike this traditional assumption, in some research, it is assumed that one job of a task is able to be executed on two or more processors concurrently and simultaneously; although, early jobs must be completed before the subsequent job can begin to be executed. This type of task is also called parallel tasks. Figure 2.4 provides a trivial example of scheduling a periodic task under this assumption. This assumption provides more space and the opportunity to use DVFS to achieve the goal of saving energy, and it makes the problem more complex and difficult at the same time.

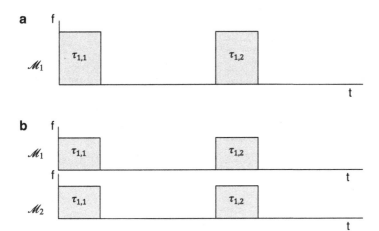

Fig. 2.4 An example of a single task being concurrently executed. (**a**) Executed on one processor. (**b**) Concurrently executed on two processors

Application-specific Power Consumption: Sometimes, it is assumed that the power consumption function of a processor is dependent on the tasks running on it, which is also the case in many practical systems. Thus, it is assumed that some constants in the power consumption model of a processor are dependent on the task(s) assigned onto it. When taking this aspect into consideration, problems become more complex than when a processor's power consumption is the same for any task.

Chapter 3
Scheduling on Homogeneous DVFS Multiprocessor Platforms

Abstract Homogeneous multiprocessor platforms are widely used on modern computing systems. Energy-aware scheduling on homogeneous platforms also receives wide research interest. This chapter surveys energy-aware scheduling research that is done on homogeneous DVFS multiprocessor platforms. These works are further classified by task types under consideration, namely, frame-based tasks, tasks with precedence constraints, periodic tasks, and sporadic tasks. Detailed techniques and algorithms are presented for various problems in the following.

3.1 Frame-Based Tasks

This section presents existing works on energy-aware scheduling algorithms for frame-based tasks on homogeneous multiprocessor platforms. Yang et al. [2] address frame-based task scheduling on dependent platforms without the consideration of task migration; Chen et al. [3] address scheduling on independent platforms with the consideration of task migration. Chen and Kuo [4] take application-specific power consumption into consideration. Kong et al. [5] considers scheduling on partitioned multi-core platforms. Li [6] makes some initial attempt to address the problem of energy-efficient scheduling of parallel tasks on multiprocessor platforms.

Scheduling on Dependent Platforms: For frame-based tasks on homogeneous dependent multiprocessor platforms, where task migration is not allowed, [2] aims to achieve scheduling with minimum energy consumption. Assuming that the number of tasks is greater than the number of processors, a schedule consists of two steps. The first step is to assign tasks to processors; the second step is to schedule the shared frequencies/speeds during different time intervals. Notice that, although the platform is dependent, which means that all processors are running at the same frequency, the shared frequency can vary over time. The active power consumption

D. Li and J. Wu, *Energy-aware Scheduling on Multiprocessor Platforms*,
SpringerBriefs in Computer Science, DOI 10.1007/978-1-4614-5224-9_3,
© The Author(s) 2013

Fig. 3.1 LTF strategy

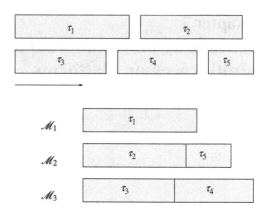

model is simplified as $P^{act}(f) = \alpha f^3$. When a processor is idle, it consumes zero power. It is proven that optimal scheduling requires the workloads assigned to processors balanced, and this assignment problem is proven to be NP-hard. The authors then adopt the Largest-Task-First (LTF) strategy to assign the task set among processors. After tasks have been assigned, without loss of generality, assuming that the utilization assigned to processor $\mathcal{M}_j$ is U_j, and $0 = U_0 \leq U_1 \leq U_2 \leq \cdots \leq U_m$, the problem of finding the optimal speed scheduling can be formulated as an optimization problem:

$$minimize \ \sum_{j=1}^{m}(m-j+1)P^{act}((U_j - U_{j-1})Df^{max}/t_j)t_j$$

$$s.t. \qquad\qquad \sum_{j=1}^{m}t_j = D$$

where t_j is the time interval between the time when $\mathcal{M}_{j-1}$ completes its workload and the time when $\mathcal{M}_j$ completes its workload. $(U_j - U_{j-1})Df^{max}/t_j$ is the frequency that should be used during time interval t_j. The above optimization problem can be solved by the Lagrange Multiplier Method. A trivial example of scheduling five tasks on three processors is provided as follows. Tasks' utilizations are: $u_1 = 0.5$, $u_2 = 0.45$, $u_3 = 0.4$, $u_4 = 0.35$, and $u_5 = 0.2$. After partitioning, utilizations assigned to different processors are: $U_1 = 0.5$, $U_2 = 0.65$, and $U_3 = 0.75$. Figure 3.1 shows the partition of the LTF strategy; Fig. 3.2 illustrates the speed scheduling after tasks are assigned.

Scheduling on Independent Platforms with Task Migration Consideration: For frame-based tasks on homogeneous independent multiprocessor platforms, when task migration is not allowed, the Largest Task First (LTF) strategy is also used to partition tasks among processors [3], and it is shown to have a good performance under various situations. When task migration and preemption are allowed, the authors propose a LTF-M strategy for the problem. The LTF-M strategy assigns each task with utilization $u_i \geq U^{total}/m$ to one processor and uses frequency $u_i f^{max}$ (in other words, use speed $s = u_i$) to execute the task such that the task completes exactly at the deadline D; then, it assigns a task with utilization $u_i < U^{total}/m$ to,

Fig. 3.2 Speed scheduling after partitioning

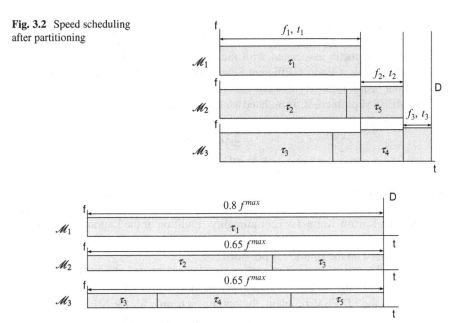

Fig. 3.3 LTF-M scheduling on three processors

at most, two processors. Denote U_{avg} as the total utilization of tasks with $u_i < U^{total}/m$ divided by the number of remaining processors (processors that have not been allocated to tasks with utilizations $u_i \geq U^{total}/m$). Using the LTF strategy, the next task with utilization $u_j < U^{total}/m$ is considered and is assigned to a processor $\mathscr{M}^*$ at frequency $U_{avg}f^{max}$. If after adding τ_j to $\mathscr{M}^*$, τ_j can meet its deadline, τ_j is added to processor $\mathscr{M}^*$; if τ_j can't meet its deadline, then the latter part of τ_j is selected to be executed on processor $\mathscr{M}^*$ at frequency $U_{avg}f^{max}$, and it will be completed exactly at deadline D; the former part of it will be executed on another processor. In practical terms, τ_j begins executing on a processor; after a certain time, it is preempted by another task. After an additional period of time, it migrates to another processor to finish its latter part.

Figure 3.3 provides a simple example that shows how the LTF-M strategy works. $u_1 = 0.8$, $u_2 = 0.4$, $u_3 = 0.4$, $u_4 = 0.3$, $u_5 = 0.2$, and $m = 3$. Since $U^{total}/m = 0.7$, $u_1 \geq U^{total}/m$, firstly, τ_1 is assigned to processor $\mathscr{M}_1$ to be executed at frequency $0.8f^{max}$. Then, $U_{avg} = (u_2 + u_3 + u_4 + u_5)/(3-1) = 0.65$. Next, τ_2 is assigned to another processor $\mathscr{M}_2$ to be executed at $0.65f^{max}$; after that, try to assign τ_3 to $\mathscr{M}_2$. Unfortunately, τ_3 can't be finished on $\mathscr{M}_2$; thus, the latter part of τ_3 will be executed on $\mathscr{M}_2$ and the former part of τ_3 will be executed on another processor $\mathscr{M}_3$, also at frequency $0.65f^{max}$. Finally, assign τ_4 and τ_5 to processor $\mathscr{M}_3$.

Application-specific Power Consumption: For a similar problem, the authors in [4] consider scheduling tasks on homogeneous independent platforms where a processors' power consumption is dependent on the task(s) running on it. Task τ_i's active power consumption function is phrased as $P^{act_i} = \alpha_i f^\varepsilon$, where ε is a hardware-dependent factor, which is regarded as between 2 and 3 in the paper, and α_i is a parameter associated with the specific task τ_i. The authors consider both Multiprocessor Energy-Efficient Scheduling with task Migration (MEESM) and without task Migration (MEES).

The MEESM problem is formulated as an optimization problem:

$$minimize \qquad \sum_{\tau_i \in \mathscr{T}} E_i(t_i^e)$$
$$s.t. \ \sum_{\tau_i \in \mathscr{T}} t_i^e = mD, 0 < t_i^e \leq D, \forall \tau_i \in \mathscr{T}$$

where t_i^e represents the execution time of τ_i, and $E_i(t_i^e)$ represents the energy consumption if executing τ_i in time t_i^e. This optimization problem can be solved by either the Karush-Kuhn-Tucker optimality condition or the Lagrange Multiplier Method. The solution is the optimal execution times $(t_1^{e*}, \cdots, t_n^{e*})$. Then, an optimal algorithm based on the solution is provided.

Denote t_i^{e*} as the estimated execution time of task τ_i. For the MEES problem, since the disallowance of task migration places more constraints on task scheduling compared to the MEESM problem, the optimal schedule for the MEESM problem produces a lower bound for the MEES problem. The authors provide a feasible schedule for the MEES problem, where task partitioning adopts the Largest Estimated Execution Time first (LEET) strategy (similar to the LTF strategy), and they prove that its approximation ratio is 1.412 by referring to the optimal schedule of the MEESM problem.

Scheduling on Partitioned Multi-core Platforms: Kong et al. [5] also consider energy-aware scheduling of frame-based tasks. The platform under consideration is a symmetric cluster-based/partitioned multi-core, where, $N_b \times N_c$ homogeneous cores are partitioned into N_b islands/blocks, each of which contains N_c cores. Cores on the same island operate at a same frequency, while cores from different islands may operate at different frequencies. Each island can adjust the frequency/voltage independently. The dynamic power consumption on a core is phrased as $P^{dyn} = \alpha f^3$. The dynamic power consumption of an island is the sum of the dynamic power of all active cores, while the static power consumption is a constant P^{sta} if at least one core in the island is active.

Consider all cores $\mathscr{M}_1, \mathscr{M}_2, \cdots, \mathscr{M}_{N_c}$ on a single island; they have already been assigned workloads $U_1, U_2, \cdots, U_{N_c}$, with $U_1 \leq U_2 \leq, \cdots, \leq U_{N_c}$. The total energy consumption can be computed as $E^{total} = \sum_{j=1}^{N_c}(E_j^{dyn} + E_j^{sta})$, where E_j^{dyn} and E_j^{sta} represent the dynamic and static energy consumption during time interval t_j, respectively (t_j is still the time interval between the time when $\mathscr{M}_{j-1}$ completes its workload and the time when $\mathscr{M}_j$ completes its workload). When there are no other constraints, by the similar method in [2], the optimal frequency setting to minimize E^{total} can be obtained: $f_j^* = \sqrt[3]{\frac{P^{sta}}{2\alpha(N_c-j+1)}}$. $\{f_1^*, \cdots, f_{N_c}^*\}$ is called the critical speed sequence (if $f_j^* > f^{max}$, let $f_j^* = f^{max}$; if $f_j^* < f^{min}$, let $f_j^* = f^{min}$).

When considering the deadline constraints, the critical speed sequence may not be optimal. A constrained convex programming problem can be formulated to minimize the energy consumption of each island:

$$minimize \quad E^{total} = \sum_{j=1}^{N_c} E_j(t_j)$$

$$s.t. \ \sum_{j=1}^{N_c} t_j \le D, t_j^{min} \le t_j \le t_j^{max}$$

where $t_j^{min} = (U_j - U_{j-1})D$, $t_j^{max} = (U_j - U_{j-1})Df^{max}/f^{min}$. Algorithm Binary Search (BS) is proposed to achieve the optimal energy consumption for this practical problem.

Due to non-negligible leakage powers and operating frequency constraints in various islands, mapping a task set to all islands will not always result in minimal energy consumption. As for determining the proper number of active islands, several steps are adopted in the paper: (a) Determine the lower bound of the number of islands required to complete the task set before the deadline, $n_b^l = \sum_{i=1}^n C_i/(N_c D)$, and the upper bound $n_b^u = min(\lceil n/N_c \rceil, N_b)$, where n is still the number of tasks. (b) Perform a linear search in the interval $[n_b^l, n_b^u]$ to determine the proper number of active islands. For each $n_b \in [n_b^l, n_b^u]$, the LTF strategy is used to partition the task set onto $n_b N_c$ cores. Then, Algorithm BS is used to determine the local minimal energy consumption for the partition of this iteration. (c) The overall algorithm finally returns the task schedule (including the number of active islands, task partition, and frequency scheduling), which results in the minimal energy value among all of the $(n_b^u - n_b^l + 1)$ iterations.

Single Task Concurrent Execution (Parallel Task Execution): In [6], the author studies the scheduling of parallel frame-based tasks on multiprocessor platforms with the consideration of energy consumption. A three-level energy/time/power allocation scheme is adopted for a given schedule for two problems, namely, minimizing the schedule length under a total energy consumption constraint and minimizing the overall energy consumption without missing any deadlines. Active power consumption P^{act} on each processor is assumed to be s^ε, where s is the processing speed and $\varepsilon(\ge 3)$ is a constant. A parallel task τ_i considered in [6] requires π_i processors to execute it concurrently, and π_i is defined as the size of task τ_i. The execution requirement of task τ_i is represented by C_i (the execution time at the maximum frequency), which is the maximum execution requirement on the π_i processors. Thus, the execution time of task τ_i, if executed at speed s_i, can be calculated as $t_i = C_i/s_i$. Notice that π_i processors need to execute task τ_i simultaneously. The total energy consumption for task τ_i can be derived: $e_i = \pi_i s_i^\varepsilon t_i = \pi_i C_i s_i^{\varepsilon-1}$.

As we have indicated, the author studies two problems: minimizing the schedule length under the total energy consumption constraint and minimizing the overall energy consumption without missing any deadlines. However, the scheduling strategy and analysis process for these two problems are quite similar to each other. Here, we only present the work for solving the first problem: minimizing the schedule length under the overall energy consumption constraint.

For the ease of analysis and scheduling, the author adopts a system partitioning approach, which partitions the m processors into $j(\geq 1)$ cluster(s) and partitions the n tasks into j group(s) such that each cluster of processors executes one group of tasks. Given a partition, the optimal frequency setting for each task can be analytically derived.

- If $\pi_i > m/2$, $\forall 1 < i < n$, any two tasks cannot be executed simultaneously for the lack of processors; in other words, all of the n tasks have to be executed sequentially. Under this situation, the optimal frequency setting for each task can be achieved.
- If $\pi_i \leq m/j$, $\forall 1 < i < n$, then the m processors are partitioned into j clusters. The n tasks are partitioned into j groups. Thus, tasks in group k can be executed on cluster k, $1 \leq k \leq j$. Notice that all tasks in one group are still executed sequentially, while tasks from different groups can be executed on different clusters simultaneously. Given a partition, the optimal power allocation for each group can also be achieved to minimize the overall schedule length. Consequently, the optimal frequency setting for each task can also be derived.

For the general case, π_i values may range randomly between 1 and m; by the partitioning rule, the system can only be considered as one cluster. Obviously, it is not efficient, because, when small-sized tasks are executed, a large number of processors remain idle, which is a waste of time and energy. The author comes up with a dynamic harmonic scheme to address the system partitioning and task scheduling problem. The author proposes a scheme which divides the original list of n tasks into c sub-lists (c is determined according to task sizes). For the jth sub-list, where $1 \leq j \leq c-1$, it contains tasks of sizes within $(m/(j+1), m/j]$. For sub-list c, it contains tasks of sizes within $(0, m/c]$. After this division, to schedule the jth sub-list, all of the processors are partitioned into j clusters; tasks in the jth sub-list are partitioned into the j clusters by a given List Scheduling (LS) algorithm, and the c sub-lists are considered one by one.

This harmonic system partitioning and task scheduling scheme performs well because tasks in one sub-list have similar sizes, which is claimed to guarantee that processor utilization will be kept at a high level. Given a partition scheme of scheduling a task group into clusters, the optimal supply voltage and frequency setting can be obtained by the previous results.

We will give an example to show how all of the harmonic system partitioning and task scheduling processes work. Consider scheduling 20 tasks on 12 processors, where $\pi_1 = 10$, $\pi_2 = 12$, $\pi_3 = \pi_5 = 5$, $\pi_4 = \pi_6 = 6$, $\pi_7 = \pi_8 = \pi_9 = \pi_{10} = \pi_{11} = \pi_{12} = 4$, and $\pi_{13} = \pi_{14} = \pi_{15} = \pi_{16} = \pi_{17} = \pi_{18} = \pi_{19} = \pi_{20} = 3$. Since the minimum size of all tasks is 3, we need to divide the list into $12/3 = 4$ sub-lists at most; in other words, $c = 4$. According to the process, sub-lists 1, 2, 3 and 4 contain tasks with sizes within $(6, 12]$, $(4, 6]$, $(3, 4]$ and $(0, 3]$, respectively. Assume that tasks within the same sub-list are already in descending order of their execution requirement and that the list scheduling algorithm adopts the largest requirement first strategy. For sub-list 1, all of the 12 processors are considered as 1 cluster; for sub-list 2, all of the

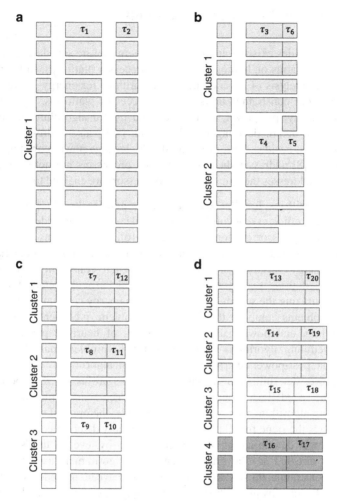

Fig. 3.4 Harmonic system partitioning and task scheduling. (**a**) Scheduling sub-list 1. (**b**) Scheduling sub-list 2. (**c**) Scheduling sub-list 3. (**d**) Scheduling sub-list 4

12 processors are partitioned into 2 clusters; the similar partitions apply to sub-list 3 and sub-list 4. The overall system partitioning and task scheduling are shown by Fig. 3.4a–d. Given this system partitioning and task scheduling, the optimal energy allocation for each task sub-list and the optimal frequency setting for each task can be determined, according to previously proven results.

3.2 Tasks with Precedence Constraints

This section provides existing approaches for energy-aware scheduling tasks with precedence constraints on homogeneous multiprocessor platforms. Li [7] provides the design and analysis of several heuristic algorithms for energy-aware scheduling of precedence-constrained tasks on multiprocessor platforms. Three types of algorithms, namely, pre-power, post-power, and hybrid algorithms are provided and analyzed in detail. de Langen and Juurlink [8, 9] also address precedence-constrained tasks; besides, both of them take the leakage power consumption into consideration. [8] addresses the situation when processors are not enabled to be shut down. [9] addresses the situation when processors are enabled to be shut down.

Heuristic Algorithms for Precedence-constrained Tasks: In [7], the author considers energy efficient scheduling of sequential tasks with precedence constraints on DVFS multiprocessor platforms. Similar to [6], two problems are addressed: minimizing the schedule length under an energy consumption constraint and minimizing energy consumption with a schedule length constraint. Again, because the similarity of these two problems, we will only provide details for the first problem: minimizing the schedule length under an energy consumption constraint. Three sub-problems compose the scheduling problem: precedence constraining, task scheduling, and power supply. Three types of heuristic power allocation and task scheduling algorithms are proposed for the problem, namely, pre-power determination, post-power determination, and hybrid algorithms.

In pre-power determination algorithms, all tasks are assumed to be executed with the same speed. Task scheduling and precedence constraining are dealt by a list scheduling algorithm; after achieving a schedule, the optimal shared execution speed (which is the same for all tasks) can be optimally determined.

In post-power determination algorithms, a level-by-level (a task's level is its depth in the directed acyclic graph) scheduling algorithm is used to deal with precedence constraints; thus, tasks in the same level have no precedence constraints, and can be scheduled by an arbitrary list scheduling algorithm; under this situation, given a task schedule, the optimal power supply for each level and each task can be optimally determined.

In hybrid algorithms, precedence constraints are also dealt with by a level-by-level scheduling algorithm. Besides, hybrid algorithms further assume that tasks at the same level should be executed at the same speed, although tasks at different levels may be executed at different speeds. Under this situation, given a task schedule, the optimal power supply for each level and each task can also be optimally determined.

Consider the following example of scheduling seven tasks on three processors. Tasks' precedence constraints are shown in Fig. 3.5a. Their execution requirements are as follows: $C_1 = 2$, $C_2 = 5$, $C_3 = 6$, $C_4 = 1$, $C_5 = 3$, $C_6 = 7$, $C_7 = 4$. Assume that when a list-scheduling algorithm is needed, the largest requirement first strategy is adopted. In the pre-power algorithm, precedence constraining and task

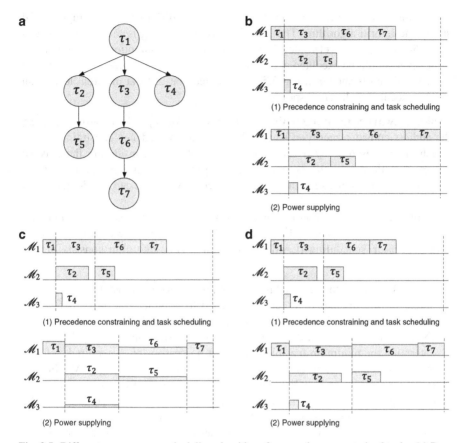

Fig. 3.5 Different energy-aware scheduling algorithms for precedence-constrained tasks. (**a**) Precedence constraints. (**b**) Pre-power algorithm. (**c**) Post-power algorithm. (**d**) Hybrid algorithm

scheduling are handled together by a list scheduling that incorporates the precedence constraints. Figure 3.5b shows the pre-power algorithm. Figure 3.5b (1) gives the original schedule. Under this scheduling and the assumption that all tasks should be executed at the same speed, the optimal power supply can be determined, where a speed as high as possible is adopted as long as the overall energy consumption is less than or equal to the energy consumption constraint. Notice that we are discussing minimizing the scheduling length here. Figure 3.5c shows the post-power algorithm. From the precedence constraints in Fig. 3.5a, we can easily notice that τ_1 is at the first level, τ_2, τ_3, τ_4 are at the second level, τ_5, τ_6 are at the third level, and τ_7 is at the fourth level. Figure 3.5c (1) shows the level-by-level scheduling. Under this schedule, the optimal power supply can be determined and is shown in Fig. 3.5c (2).

Notice that tasks at the same level can execute at different speeds. Figure 3.5d shows the hybrid algorithm. The level-by-level scheduling is the same as that in Fig. 3.5c (1). Under level-by-level scheduling and the assumption that tasks at the same level should be executed at the same speed, the optimal power supply can be determined and is shown in Fig. 3.5d (2). Notice that tasks at different levels can still be executed at different speeds.

Leakage-aware Scheduling without Shutdown: For dependent platforms, in [8], the authors consider energy-efficient scheduling for precedence-constrained tasks, which are represented by a task graph model. Without considering the leakage power or static power, the optimal scheduling involves assigning the tasks to as many cores as possible, such that a minimum common speed can be adopted on all of the cores.

While static power is non-negligible, using as many cores as possible may not be the optimal solution. For this problem, the authors provide a Leakage-Aware Multiprocessor Scheduling (LAMPS) algorithm. This algorithm consists of two steps.

- Step 1: Determine the minimal number of processors. First, determine the lower bound on the number of processors needed to complete the tasks before the deadline, which is simply the overall workload divided by the deadline: $N_l = \lceil \sum_{\tau_i \in \tau} C_i / D \rceil$, as well as the upper bound $N_u = n$ (the number of tasks). Then, determine the minimum feasible number of processors, N_{min}, required to finish the task set on time. This is achieved by a binary search between $[N_l, N_u]$. First, it is determined if $N = (N_l + N_u)/2$ will finish before the deadline. A list scheduling that employs the Earliest Deadline First (EDF) scheme is used to check the feasibility. If the makespan of the schedule produced by the list scheduler satisfies the deadline constraints, the search continues on the interval $[N_l, N]$; otherwise, the search continues on the interval $[N + 1, N_u]$.

- Step 2: Determine the number of processors that requires the least amount of energy. This step also consists of two steps. First, determine the total power consumption for N_{min} processors. This is done by lowering the clock frequency and supply voltage so that the task set is completed exactly at the deadline. This is also done for $N_{min} + 1$, $N_{min} + 2$, etc., processors, until increasing the number of processors no longer decreases the makespan of the schedule.

Finally, the optimal number of processors that requires the least amount of energy is determined.

Leakage-aware Scheduling with Shutdown: In the above work [8], though leakage power is considered to achieve energy efficient scheduling, it is assumed that the processors are unable to be shutdown, even when they are idle. If processors are able to be shutdown, energy consumption can be lowered further. In [9] the authors propose a Leakage-Aware Multiprocessor Scheduling heuristic with the option to Shutdown Processors, called LAMPS+PS, which provides tradeoffs between three techniques: DVFS, processor shutdown, and finding the optimal number of processors. The number of processors that minimizes the energy consumption is determined by calculating the energy consumption when using N_{min} processors, $N_{min} + 1$ processors, until the makespan will not be reduced, even by using more processors. For each number of processors, the tradeoff between DVFS and processor shutdown is determined by scaling the frequency from the maximum to the minimum frequency (which is required to meet the deadline) to achieve the minimal energy consumption. Finally, the optimal number of processors that requires the least amount of energy is obtained, as well as the corresponding optimal tradeoff between DVFS and processor shutdown.

3.3 Periodic Tasks

This section surveys the state-of-the-art research that focuses on energy-aware scheduling of periodic tasks on homogeneous multiprocessor platforms. Aydin and Yang [10] address the problem of power-aware partitioning of periodic tasks among multiple processors. Chen and Kuo [11] consider application-specific power consumption. Chen et al. [12] take the leakage power consumption into consideration. Xian et al. [13] consider the scheduling problem when probability distributions of tasks' execution requirements are known before scheduling. Lee [14] assumes that one job from a task can be concurrently executed on multiple processors and deals with the energy-aware scheduling under this assumption. Taking into consideration the leakage power consumption and application-specific power consumption, [15] addresses the problem of determining the optimal number of cores to use, and after that, provide two online schemes to reduce energy consumption further: slack reclamation and load refining. Seo et al. [16] provide two schemes to dynamically determine the number of active cores and to migrate tasks during runtime. Zhang et al. [17] provide an optimal energy-efficient global scheduling scheme, which requires frequent task migration and preemption. Zeng et al. [18] provide a scheduling algorithm (for both Earliest Deadline First (EDF) and Rate Monotonic (RM) scheduling), which takes several practical constraints into consideration.

Partition-based Scheduling without Migration: In [10], the authors consider the problem of partitioning periodic real-time tasks on a homogeneous multiprocessor platform by considering both feasibility and saving energy, where task migration is not allowed. The objective is to derive a feasible partitioning policy that results in minimum energy consumption, while meeting all timing requirements with dynamic priority EDF scheduling. Firstly, it is proven that this problem is NP-hard. It is then shown that a task partition that evenly divides the total workload among all of the processors, if it exists, will minimize the total energy consumption.

There are several heuristics for the partition problem: First-Fit (FF), Best-Fit (BF), Next-Fit (NF), and Worst-Fit (WF). The authors provide heuristics for this problem when either utilization ordering is known a priori or not. It is proven that when utilization ordering is known a priori, the Worst-Fit Decreasing (WFD) heuristic always achieves balanced partitioning and thus optimal energy conservation. When utilization ordering is not known before scheduling, it is observed that there doesn't exist a clear winner that performs well in all utilization values in terms of both feasibility and energy consumption among FF, NF, BF and WF. It has also been observed that FF, NF, and BF offer good performance in terms of feasibility, while WF offers a good performance in terms of energy consumption. Then, the authors provide an algorithm called RESERVATION to tackle this problem, which is a trade-off between the good feasibility performance of FF, BF and the good energy conservation of WF. The idea of RESERVATION is basically to reserve $\lfloor m/2 \rfloor$ processors for "light" tasks with utilization $u_i \leq U^{total}/m$ and allocate the remaining processors for other tasks with utilization $u_i > U^{total}/m$.

Application-specific Power Consumption: The authors in [11] address energy-efficient multiprocessor scheduling of periodic real-time tasks with different power consumption functions, where task migration is not allowed. The active power consumption function of task τ_i is phrased as $P^{act_i} = \alpha_i f^{\varepsilon}$, where α_i is a parameter related to the specific task, and $\varepsilon \leq 3$. Firstly, the minimization problem of the energy consumption for multiprocessor scheduling considered in the paper is proven to be NP-hard, and is reformulated as a convex optimization problem. By allowing tasks to migrate to different processors, the convex optimization problem can be relaxed to:

$$minimize \qquad \sum_{\tau_i \in \mathscr{T}} E_i(t_i^e)$$
$$s.t. \ \sum_{\tau_i \in \mathscr{T}} t_i^e/T_i = m, 0 < t_i^e \leq T_i$$

where t_i^e represents the execution time of a job of task τ_i, $E_i(t_i^e)$ is the amount of energy consumed by all of the jobs of τ_i during the hyper-period H, if executing at frequency $WCEC_i/t_i^e$. By applying the Karush-Kuhn-Tucker optimization condition, the optimal solution is achieved: $(t_1^{e*}, t_2^{e*}, \cdots, t_n^{e*})$. Let $u_i^* = t_i^{e*}/T_i$ be called the estimated utilization of τ_i. A Largest Estimated Utilization First (LEUF) strategy is proposed for the practical problem without task migration and is denoted by Algorithm LEUF, which is proven to have an approximation ratio of $\frac{(\varepsilon-1)^{\varepsilon-1}(2^{\varepsilon}-1)^{\varepsilon}}{\varepsilon^{\varepsilon}(2^{\varepsilon}-2)^{\varepsilon-1}}$.

Leakage-aware Scheduling on Independent Platforms: Chen et al. [12] address energy-efficient scheduling of periodic real-time tasks on ideal homogeneous independent platforms, while considering the consideration of leakage power consumption. The active power consumption of a processor is phrased as $P^{act} = f^3 + \beta$. It is obvious that the power consumption is a convex and increasing function of f, while the energy consumption per cycle $E^{act} = f^2 + \beta/f$ is just a convex function of f and has a local minima, where the frequency is called critical frequency, denoted by f^{crit}. The Largest Task First is also applied for problems considered in this paper. If the energy switching overhead E_j^{ov} is negligible, using all off the processors will not increase the switching overhead. Thus, after achieving a partition, if the lowest required frequency of the workload on a processor is smaller than the critical frequency, the scheduler executes the tasks on that processor at the critical speed. It is shown that the above algorithm, denoted by Algorithm LA+LTF (Leakage-Aware + Largest Task First), is a 1.283-approximation algorithm.

It can be noted that adopting the above process will result in a schedule where multiple processors execute at the critical frequency. If the energy switching overhead is non-negligible, the produced schedule will consume a significant amount of switching energy. Thus, after achieving a partition by the LTF strategy, efforts should be made to reduce the number of applied processors. The following approach is applied to consolidate workloads on the processors that are executing at the critical frequency.

- In the first step, collect all of the tasks that are executed at the critical frequency, denoted by task set $\mathcal{T}^*$. Suppose that m^* is the number of processors that execute tasks at the critical speed after the LA+LTF (Notice that the other $m - m^*$ processors are operating at a frequency greater than f^{crit}).
- Then, reassign tasks in $\mathcal{T}^*$ to these m^* processors (possibly a subset of these processors) by applying the first-fit algorithm with a bin height of f^{crit}/f^{max}. Firstly, mark the m^* processors as unused and the tasks in $\mathcal{T}^*$ as unassigned. Then, assign an unassigned task to the first-used processor if, after assigning, the accumulated utilization on this processor is still not greater than f^{crit}/f^{max}; if this task can't be assigned to any used processor, such that the accumulated utilization is not greater than f^{crit}/f^{max}, it will use another processor in the group of m^* processors.

After these steps, the EDF strategy is adopted in each processor to meet all of the tasks' timing requirements. This algorithm is denoted by Algorithm LA+LTF+FF (Leakage-Aware + Largest Task First + First Fit), which is proven to be a 2-approximation algorithm.

Consider the following example as shown in Fig. 3.6; after LA+LTF, $U_1 = 0.8$, $U_2 = 0.4$, $U_3 = 0.2$, assuming that $f^{crit} = 0.7f^{max}$, then $\mathcal{M}_1$ will be executing at $0.8f^{max}$. $\mathcal{M}_2$ and $\mathcal{M}_3$ will be executing at f^{crit}. By the LA+LTF+FF algorithm, workloads on $\mathcal{M}_2$ and $\mathcal{M}_3$ can be consolidated onto $\mathcal{M}_2$. Thus, the total energy consumption is further reduced.

Scheduling Based on Probability Distributions of Tasks' WCETs: The authors in [13] also consider energy-aware scheduling for periodic tasks on homogeneous platforms. The main difference and contribution is that they consider the Probabilistic Distribution Functions (PDFs) of the tasks' execution times to partition the workload for energy conservation. Their consideration and method is based on the previous work of [19], where the probabilistic distributions of the tasks' execution times is considered for energy-aware scheduling on a single processor, and the optimal frequency scheduling is also provided. Consider the PDFs of tasks; though their WCETs may be equal, their expected workloads can still be quite different. Hence, to balance the workload partition among processors, based on the WCETs, is not optimal. Instead, the authors aim to achieve the balanced expected workload partition via the PDFs of tasks' execution times. In [19], the range of $[0, C_i]$ is

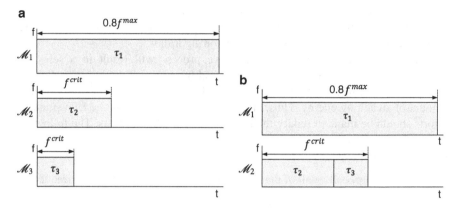

Fig. 3.6 LA+LTF VS LA+LTF+FF. (**a**) Scheduling by LA+LTF. (**b**) Scheduling by LA+LTF+FF

divided into l_i bins; each bin contains an equal amount of execution time: $e_i = C_i/l_i$; then, the distribution of a task's execution time is expressed by the Cumulative Distribution Function (CDF), denoted by Ψ. Thus, the probability of τ_i consuming the jth bin of execution time is $1 - \Psi_i(j-1)$. Obtaining the optimal frequency scheduling requires knowing how to assign frequency to the jth bin for task τ_i, denoted by $f_{i,j}$, such that the overall energy consumption is minimized. Represent the expected energy consumption in terms of Ψ and $f_{i,j}$. The optimal frequency scheduling is achieved in [19].

Xian et al. [13] extend this approach onto multiprocessor platforms. Their main contribution is that they provide balanced partitioning in terms of expected execution time. Consider scheduling four tasks, $\tau_1, \tau_2, \tau_3, \tau_4$, on two identical processors, all of which have the same WCET of 2 s. However, τ_1 and τ_2 are identical and always consume this WCET; τ_3 and τ_4 are identical, and the probability to consume 1 s is 80 %, while the probability to consume 2 s is 20 %. If we are simply using tasks' WCETs to partition the task set, one *optimal* partition would require τ_1, τ_2 to one processor and τ_3, τ_4 to another. However, this partition is not optimal in terms of the expected execution requirement since, in most circumstances, τ_3, τ_4's execution times will be less than 2 s. Instead, assigning τ_1, τ_3 to one processor and τ_2, τ_4 to the other is the optimal partition.

Assuming ideal processors with unbounded frequencies, the formulated expected energy minimization problem can be transformed into a load balancing problem. With practical values of tasks' execution times, the workload cannot be exactly balanced. Thus, the Worst-Fit Decreasing (WFD) algorithm is used, for WFD has been shown to be the best bin-packing heuristic for balancing loads among multiple bins [10]. After partitioning by WFD, the optimal ideal frequency scheduling is achieved by the results in [19]. However, practical processors have bounded discrete frequencies. Then, a series of procedures is applied to restrict the frequencies within the set $\{f^{min}, \cdots, f^{max}\}$, while still meeting all of the tasks' deadlines and achieving minimal energy consumption.

Single Task Concurrent Execution (Parallel Task Execution): Lee [14] addresses an energy-efficient scheduling scheme of periodic real-time tasks on lightly loaded multi-core platforms, where the number of cores is greater than the number of running tasks. Though migration is not allowed, it is assumed that the periodic tasks are able to be executed concurrently on more than one core. The authors first prove that this problem is NP-hard, and then propose a heuristic algorithm to find an energy-efficient scheduling. Since static power consumption is non-negligible, it is not optimal to use all of the cores. For a single core, the average power consumption is determined by the utilization assigned to it. A threshold utilization is defined as Uth, such that $P(2U$th$) = 2P(U$th$)$. Based on this utilization threshold, the given tasks are classified into three categories: Heavy Tasks with $u_i \geq 2U$th, Medium Tasks with Uth $\leq u_i \leq 2U$th, and Light Tasks with $u_i \leq U$th. Based on the convexity and increasing property of $P(U)$, it is derived that: assigning two Light Tasks to a single core consumes less energy than assigning them to two separate cores; if the accumulated utilization of a Light Task and a Medium Task is no larger than $2U$th, assigning the two tasks to a single core consumes less energy than assigning them to two separate cores, and vice versa; assigning two Medium or Heavy Tasks on two separate cores consumes less energy than assigning them to a single core.

If the multi-core platform has sufficient cores, the algorithm may assign tasks to as many cores as possible. For each Heavy Task, the algorithm calculates the total power consumption of executions on all possible numbers of cores, i.e., $P(U_m^\eta)$ for $1 \leq \eta \leq B_i$, where B_i is the maximal number of cores that task τ_i can be concurrently executed on. Denote the value of η, which minimizes $P(U_i^\eta)$, by η_i. Task τ_i will be assigned to η_i cores. After that, Medium Tasks and Light Tasks are assigned to the remaining cores (that are not used by Heavy Tasks). The First-Fit Decreasing (FFD) heuristic, with a bin height of $2U$th, is applied. The FFD sorts tasks in descending order of utilization and assigns them one by one to the first core whose accumulated utilization after assigning is not greater than $2U$th.

If there are less available cores than, required by the above approach, the Worst-Fit Decreasing (WFD) heuristic is used. Given m cores, the algorithm divides them into m_1 and m_2 cores, such that $m_1 + m_2 = m$, and $m_1 \geq m_h$, where m_h is the number of Heavy Tasks. First, it searches for the best assignment of the m_1 cores, where the total energy consumption of all Heavy Tasks is minimized. Next, it assigns Medium Tasks and Light Tasks to m cores according to the WFD heuristic. The total energy consumption of the m cores for each combination of $m_1 + m_2 = m$ is calculated, and finally, the best combination is selected.

Consider the following example: $u_1 = 0.8$, $u_2 = 0.4$, $u_3 = 0.2$. Uth $= 0.225$. Thus, τ_1 is a Heavy Task, τ_2 is Medium Task, and τ_3 is a Light Task. Assume that the optimal number of processors to execute τ_1 concurrently is 3. After that, $U_1 = U_2 = U_3 = 0.4$ are the utilizations of τ_1 split onto $\mathcal{M}_1, \mathcal{M}_2, \mathcal{M}_3$, respectively.

Fig. 3.7 Assignment of sufficient cores

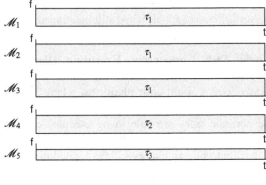

Fig. 3.8 One possible assignment of insufficient cores

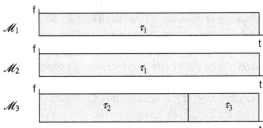

Note that $U_1 + U_2 + U_3$ can be greater than u_1, for concurrent execution may call for some overhead. Since $u_2 + u_3 = 0.6 > 2U$th $= 0.45$, τ_2 and τ_3 will be assigned to separate processors. The final schedule is shown in Fig. 3.7.

If there are only three processors (less than that five that are used in the case of sufficient cores), then there are three combinations of $(m_1 + m_2) = 3$, since m_1 can be 1,2, or 3. Suppose that the optimal combination is $(m_1, m_2) = (2, 1)$. After assigning the Heavy Task τ_1 to $\mathcal{M}_1$ and $\mathcal{M}_2$, $U_1 = U_2 = 0.5$, which are the utilizations of τ_1 split onto $\mathcal{M}_1$ and $\mathcal{M}_2$, respectively. Notice again that $U_1 + U_2$ can be greater than u_1. τ_2 and τ_3 are assigned by the WFD strategy. One possible optimal assignment of insufficient cores is shown in Fig. 3.8.

Scheduling on Dependent Platforms with Dynamic Coordination: The authors in [15] undertake the problem of scheduling periodic tasks on dependent Chip Multi-core Processors (CMPs) with the objective of minimizing the overall energy consumption. The power consumption required to execute τ_i on processor $\mathcal{M}_j$ is modeled as: $P_j^{act} = P^{sta} + a_i f^3 + P^{ind_i}$, where a_i and P^{ind_i} are dependent on the tasks assigned to the processor. Based on the power model, a global energy-efficient frequency threshold for k active cores is derived: $f_{ee}(t) = \sqrt[3]{\frac{P^{ind}(t)}{2a(t)}}$, where $P^{ind}(t) = \sum_{i=1}^{k} P^{ind_i}$ and $a(t) = \sum_{i=1}^{k} a_i$. $f_{ee}(t)$ is similar to the critical frequency for a uniprocessor, and this means that a scheduling operating frequency under $f_{ee}(t)$ will not result in energy conservation; in other words, as long as the processor is on, it should operate at a frequency no less then $f_{ee}(t)$ if it aims to consume less energy.

For a given task set, the first step is to choose the number of active cores and to conduct task partitioning. Three algorithms are proposed: Sequential Search (SS) Algorithm, Greedy Load Balancing (BLB) Algorithm, and Threshold-based Load Balancing Algorithm. Algorithm SS exhaustively searches from the minimum number of necessary cores, $\lceil U^{total} \rceil$, to m. For each $k \in [\lceil U^{total} \rceil, m]$, it generates a partition $\mathscr{P}_k$ using WFD. It computes the expected energy consumption of the feasible partition $\mathscr{P}_k$. The k value with the least expected energy is returned. Algorithm GLB invokes WFD once on all m cores. After partitioning, GLB tries to move all tasks from the least loaded core to the second least loaded core, if and only if feasibility is still guaranteed and the expected energy consumption is not increased after moving these tasks. The algorithm iterates for the remaining cores until it is no longer possible to conduct such a task moving. Algorithm TLB uses the concept of load threshold, where a partition is accepted by TLB if the minimum load on any core is no smaller than a predefined threshold. TLB first invokes WFD once on all m cores and then iteratively tries to move all tasks from the core with the least load to the core with the second least load, if the least load is smaller than the threshold and doing so doesn't contradict the feasibility. After such a move, the algorithm is iteratively reinvoked on the new set of active cores.

After choosing the number of active cores and partitioning the task set to these cores, two coordinated voltage and frequency scaling approaches are proposed. Let U_j be the utilization assigned to core $\mathscr{M}_j$. $U(t) = max(U_j), j = [1, \cdots, m]$ is the largest load value among all active cores. The first approach, called CVFS, consistently sets the shared frequency $f(t) = max(U(t), f_{ee}(t))$, which is based on the static load values of active cores.

However, there are potential benefits in computing the instantaneous load U_j^*, which is addressed by CVFS*. CVFS* works for two reasons: (a) some jobs may not take their WCECs and may complete early. Due to this unused CPU time, in some intervals, the instantaneous load of $\mathscr{M}_j$ may be less than U_j; (b) due to the constraints imposed by $f_{ee}(t)$ and the global voltage/frequency, a given processor may be forced to execute at frequency levels that are higher than necessary. Hence, its remaining workload may be lower than U_j in some intervals. Corresponding to the first reason, a slack reclaiming technique is applied; corresponding to the second reason, a load-refining technique is adopted. Both techniques are implemented by dynamically setting the effective workload of task τ_i, denoted by $u_i(t)$. When a task arrives, its initial u_i is set to be C_i/T_i. In the slack reclaiming technique, if a job of task τ_i, released at time t_c, completes after executing $ACEC_i (\leq WCEC_i)$ Actual Case Execution Cycles, then the effective utilization of τ_i during interval $[t_c, t_c + T_i]$ is set to be $ACEC_i/(f^{max} T_i)$. In the load refining technique, the basic principle is: on core $\mathscr{M}_j$, the execution of a task τ_i at a frequency $U_j' f^{max} > U_j f^{max}$ may be seen as equivalent to executing a workload $ACEC_i' < WCEC_i$ at speed $U_j f^{max}$. Thus, after its completion, this core's effective workload can also be refined.

Consider the following example: $\tau_1 = (15, 20, 20)$, $\tau_2 = (3, 20, 20)$, $\tau_3 = (4, 20, 20)$, $\tau_4 = (7, 40, 40)$. Suppose that τ_2, τ_3, τ_4 will consume their WCETs, while τ_1 will only consume 3 units of its WCET. τ_1 is assigned to $\mathscr{M}_1$, τ_2 and τ_3

Fig. 3.9 Slack reclamation

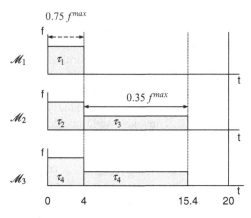

Fig. 3.10 Load refining

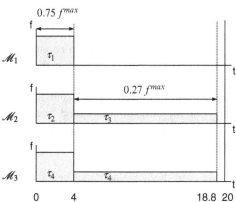

are assigned to $\mathcal{M}_2$, and τ_4 is assigned to $\mathcal{M}_3$. Utilizations of all cores are $U_1 = 0.75$, $U_2 = 0.35$, $U_1 = 0.175$. At the beginning, all cores must operate at $max(0.75, 0.35, 0.175) f^{max} = 0.75 f^{max}$. After 4 time units, τ_1 is completed. Thus, τ_1's effective utilization is changed to $3/20 = 0.15$; U_1 is also changed to 0.15. By the slack-reclamation scheme, after time 4, it is feasible to execute at frequency $max(0.15, 0.35, 0.175) f^{max} = 0.35 f^{max}$. Both τ_3 and τ_4 will complete at time 15.4, as shown in Fig. 3.9.

By the load-refining scheme, it is notable that, at time 4, τ_2 also completes; it can be regarded that τ_2 only consumes $4 * 0.35 = 1.4$ units of its WCET at $0.35 f^{max}$. Thus, the utilization of τ_2 can also updated as $1.4/20 = 0.07$. Hence, U_2 is updated as $1.4/20 + 4/20 = 0.27$. It is feasible to operate at $max(0.15, 0.27, 0.175) f^{max} = 0.27 f^{max}$ after time 4. Both τ_3 and τ_4 will complete at time 18.8, as shown in Fig. 3.10. In this way, the load-refining scheme reduces the processors' speed further, thus reducing the energy consumption further as well.

Online Core Switching and Task Migration: The authors in [16] also consider partition-based scheduling for periodic tasks. It is demonstrated that minimizing

P^{dyn} in a multi-core processor for a given task set is essentially a problem of generating the most balanced partition. However, even though the initial partitioned state can be well-balanced, the performance demand on each core may change during runtime. Thus, to achieve consistently low power consumption, the performance demand of each core must stay balanced during runtime. The intuitive way to solve this temporal imbalance is by migrating some tasks on the fly from a core with a higher workload to a core with a lower workload. The slack between tasks' actual case execution times and worst case execution times are considered when determining the most effective utilization during runtime. At every task's arrival or completion, the proposed dynamic repartitioning algorithm updates the dynamic utilization of all the active cores and repeatedly finds the cores with the maximum and minimal utilization $\mathcal{M}_{max}$ and $\mathcal{M}_{min}$, respectively. Then, migrate a task from $\mathcal{M}_{max}$ to $\mathcal{M}_{min}$, if all of the dynamic utilizations are more balanced after migration, until the whole system achieves the most balanced state.

Also, since the leakage power is assumed to be non-negligible, it is not optimal to use as many cores as possible. A dynamic core scaling algorithm is also proposed to dynamically determine the optimal number of active cores. The expected power consumption function depends on the number of active cores and the dynamic utilization of the task set. Thus, given the current dynamic utilization of the task set, the optimal number of active cores can be determined. If the number of currently active cores is less than the optimal number, additional cores are activated and a dynamic repartitioning algorithm is called to rebalance the workloads among all of the cores; otherwise, cores with low utilization will be deactivated after migrating their workloads to other cores.

Optimal Global Scheduling: In [17], Real-Time Static Voltage and Frequency Scaling (RT-SVFS) techniques are proposed for periodic tasks on homogeneous multiprocessor platforms. The techniques are regarded as static because after setting the initial frequency/speed, processors' supply voltages and execution frequencies will not change during runtime, which is considered to be better than dynamic scaling when transition overhead is significant. The authors proposed two separate techniques for when the platform can control the voltage and frequency either uniformly or independently, both of which are based on an optimal real-time scheduling algorithm for multiprocessor platforms: LLREF algorithm [20]. LLREF stands for Largest Local Remaining Execution First strategy.

In the following, we would like to introduce the LLREF algorithm first. Notice that in LLREF, task migrations and preemptions are fully allowed, as long as the same task is not executed parallel on more than one processor. The LLREF algorithm is considered to be optimal because it can schedule tasks in a way such that all tasks meet their deadlines when the total utilization demand does not exceed the utilization capacity of the platform. Formally put, given a periodic task set $\mathcal{T} = \{\tau_1, \tau_2, \cdots, \tau_n\}$, if and only if the total utilization of the task set U^{total} satisfies the condition: $U^{total} \leq m$, and $u_i \leq 1$ for all $i = 1, 2, \cdots, n$, then this task set (with implicit deadlines) can be scheduled to meet all deadlines by LLREF. The LLREF

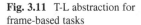

Fig. 3.11 T-L abstraction for frame-based tasks

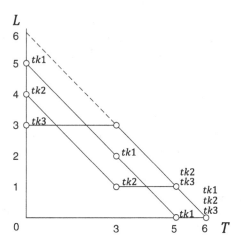

algorithm is based on the T-L plane (Time and Local Execution Time Domain Plane), where the horizontal axis represents the current time, and the vertical axis represents tasks' local remaining execution times.

In the T-L plane, each task's state is represented by a token. The token's horizontal-axis value describes the current time, and its vertical-axis value represents the task's local remaining execution time. The remaining execution time of a task here means that it must be consumed by the end of this T-L plane. Each token moves in the T-L plane, while scheduling decisions are made over time. Tokens are only allowed to move in two directions. When the task is selected and executed, its token moves down diagonally. Otherwise, it moves horizontally. If m processors are considered, at most m tokens can move down diagonally and simultaneously. The scheduling objective in the current T-L plane is to make all tokens arrive at the rightmost vertex of the T-L plane with zero local remaining execution time.

For ease of understanding, we will incorporate the T-L plane abstraction into a scheduling process of a frame-based task set first. Consider scheduling the following frame-based task set: $C_1 = 5$, $C_2 = 4$, $C_2 = 3$, on two processors, where the tasks' common deadline is 6. Obviously, there is no way that a partition-based scheduling can schedule these three tasks; however, since their total utilization is exactly 1, the LLREF algorithm can schedule them. Since the three tasks share a common deadline, we only need to construct one T-L plane, as shown in Fig. 3.11, where, tk_1, tk_2, and tk_3 represent the states of tasks τ_1, τ_2, τ_3, respectively. Initially, the remaining execution times of τ_1, τ_2, τ_3 are 5, 4, and 3. LLREF selects the tasks with the largest local remaining execution time to execute, namely τ_1 and τ_2 in this example. Thus, tk_1, tk_2 move down diagonally and tk_3 moves horizontally. After 3 units of time, tk_3 hits the No Local Laxity Diagonal (NNLD), which is called a *ceiling-hitting event* (also called event C). At this time instant, task τ_3 must be selected to execute, otherwise it will miss its deadline. Thus, tk_1 and tk_3 will move down diagonally and tK_2 will move horizontally. After an additional 2 units of time, tk_1 hits the bottom line, which is called a *bottom-hitting event* (also called event B).

Fig. 3.12 Corresponding execution of the tasks on two processors

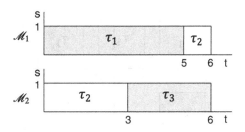

At this time instant, LLREF considers selecting another task to execute, namely tk_2 in this example. Thus, tk_2 and tk_3 will move down diagonally and tk_1 will move horizontally. At time instant 6, all tokens arrive at the rightmost point of this T-L plane. The corresponding task execution is shown in Fig. 3.12.

For periodic tasks, LLREF divides the time axis into consecutive T-L planes, such that each T-L plane ends at some task's deadline and there are no deadlines within any T-L plane. It is proven that if tasks' execution requirements in each T-L plane can be satisfied, the whole scheduling is feasible. To ensure that tasks in each T-L plane consume their execution requirement, the LLREF adopts the following strategy: in each T-L plane, initially, m of the largest local remaining execution time tasks are selected first; when the system encounters an events B or C (refer to the descriptions in the above frame-based task set example), the local remaining execution time of tasks will be updated, and another set of m largest local remaining execution time tasks are selected. In this way, local schedulability is guaranteed, and the optimality of the algorithm is proven consequently.

Consider the following example where $\tau_1 = (3,4,4)$, $\tau_2 = (3,6,6)$, $\tau_1 = (6,8,8)$. Their total utilization is 2, so they can be feasibly scheduled on two processors. Task τ_1's deadlines are 4, 8, 12, 16, 20, 24; task τ_2's deadlines are 6, 12, 18, 24; task τ_3's deadlines are 8, 16, 24. Thus, the whole time domain in one hyper-period can be divided into 8 T-L planes, with intervals (0, 4), (4, 6), (6, 8), (8, 12), (12, 16), (16, 18), (18, 20), and (20, 24), respectively, which is demonstrated in Fig. 3.13.

In the first T-L plane, within (0, 4), initially task τ_1's execution requirement is 3; task τ_2's execution requirement is $3/6*4 = 2$; task τ_3's execution requirement is $6/8*4 = 3$. τ_1 and τ_3 have the largest local remaining execution times and are selected to execute first. After 2 units of time, task τ_2 hits the No Local Laxity Diagonal (NNLD); in other words, an event C happens. Thus, task τ_2 is selected to execute. Either τ_1 or τ_3 will be executed on the other processor. Assume that task τ_3 is selected to execute, then, after one more unit of time, tk_1 hits the NNLD, and tk_3 hits the bottom line. Tasks τ_1 and τ_2 will be selected to execute on two processors. All three tokens, tk_1, tk_2, and tk_3 will arrive at the rightmost vertex of the first T-L plane. The first T-L plane is shown in Fig. 3.14. The other 7 T-L planes can be achieved similarly and are shown in Fig. 3.15. Correspondingly, we can get the optimal scheduling, which is shown in Fig. 3.16.

Based on the optimal LLREF algorithm described above, the authors in [17] propose two static energy-efficient algorithms for homogeneous multiprocessor systems. The first one is called Uniform RT-SVFS, which claims that a periodic task

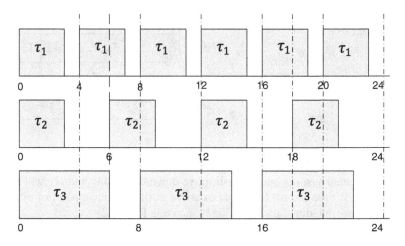

Fig. 3.13 Creating T-L planes according to periodic tasks' deadlines

Fig. 3.14 The first T-L plane

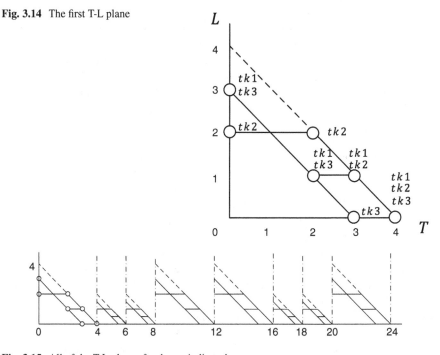

Fig. 3.15 All of the T-L planes for the periodic task set

set with total utilization $U^{total} \leq \alpha m$ and $U^{max} \leq \alpha$ will be scheduled to meet all deadlines on m processors with frequency α by LLREF. Thus, the frequency of the processors can be safely reduced to α. Notice that they assume power consumption is proportional to the cube of the processing frequency, so the overall energy

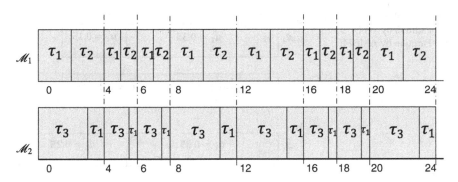

Fig. 3.16 Practical execution of the periodic tasks on two processors

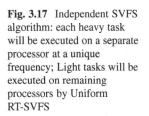

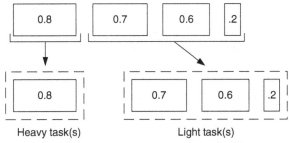

Fig. 3.17 Independent SVFS algorithm: each heavy task will be executed on a separate processor at a unique frequency; Light tasks will be executed on remaining processors by Uniform RT-SVFS

Heavy task(s)   Light task(s)

consumption is reduced. Notice that, in the first algorithm, it is required that $U^{max} \leq \alpha$, which cannot always be satisfied. If processor's execution frequencies can be scaled independently, the authors propose the second algorithm, which is called Independent RT-SVFS and does not have this requirement. The second algorithm uses the technique of classifying tasks into heavy tasks and light tasks. This technique is almost the same as that in [3]. Thus, we will not explain it again and will just give another example as shown in Fig. 3.17. Consider scheduling four tasks: $u_1 = 0.8$, $u_1 = 0.7$, $u_1 = 0.6$, $u_1 = 0.2$. By Uniform RT-SVFS, the common frequency should be set as $\alpha = (0.8 + 0.7 + 0.6 + 0.2)/3 = 0.7667$. However, since $u_1 > 0.7667$, task τ_1 can be scheduled. Uniform RT-SVFS fails here. Independent RT-SVFS detects all possible heavy tasks first. Since $u_1 > U^{total}/3$, task τ_1 is considered as a Heavy Task and will be executed on a processor with frequency $0.8 f^{max}$, or, in other words, with speed 0.8. After this, $0.7 < (0.7 + 0.6 + 0.2)/2 = 0.75$; all of the remaining tasks are considered as Light Tasks and will be executed by the first algorithm, all at frequency $0.75 f^{max}$ because the requirements of the first algorithm are satisfied considering scheduling remaining tasks on the two remaining processors now.

Practical Scheduling for Both Dynamic and Fixed Priority Assumptions: In [18], the authors target energy-efficient scheduling with practical constraints, including discrete speed, idle power, application-specific power characteristics, and inefficient speed, etc. Instead of theoretical calculation, the authors use measured

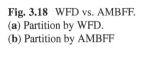

Fig. 3.18 WFD vs. AMBFF.
(**a**) Partition by WFD.
(**b**) Partition by AMBFF

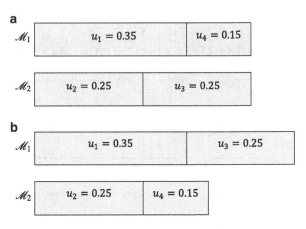

power values of $P^{act}(f)$. When the processor is in idle mode, it consumes constant P^{idl}, and the processor can only be shut down if no tasks are assigned to it. It consumes zero power in shutdown mode. An Adaptive Minimal Bound First-Fit (AMBFF) algorithm is proposed for both the Earliest Deadline First (EDF) dynamic priority and the Rate Monotonic (RM) fixed-priority scheduled periodic tasks.

Considering non-ideal processors, it is shown that Worst-Fit Decreasing (WFD), which can achieve optimal balanced workloads among multiple processors, doesn't work well for discrete speeds. Consider the following example: schedule tasks with utilizations $u_1 = 0.35$, $u_2 = 0.25$, $u_3 = 0.25$, $u_4 = 0.15$ on two processors with frequency set: $\{0.15 f^{max}, 0.4 f^{max}, 0.6 f^{max}, 0.8 f^{max}, f^{max}\}$. WFD strategy will assign τ_1 and τ_4 to one processor, and τ_2 and τ_3 to another, as in Fig. 3.18a. Though the workload assigned to each processor is 0.5, there is no speed $0.5 f^{max}$, so both processors have to operate at $0.6 f^{max}$. The proposed AMBFF scheme assigns tasks as in Fig. 3.18b. We can see that only one processor needs to operate at $0.6 f^{max}$, the other one can operate at $0.4 f^{max}$. Thus, scheduling in Fig. 3.18b is better than that in Fig. 3.18a, in terms of energy conservation.

The AMBFF algorithm works as follows: firstly, set the lowest frequency level as the frequency upper-bound for all processors; under the bound, First-Fit Decreasing (FFD) is applied; if a task cannot fit into any processor under this bound, a subsequent higher frequency is chosen to be the new bound. For the example in Fig. 3.18b, firstly, $0.15 f^{max}$ is set as the bound, then $0.4 f^{max}$, and thus τ_1 and τ_2 are assigned to two processors. Consider τ_3: it cannot be assigned to either processor under the $0.4 f^{max}$ bound; thus, $0.6 f^{max}$ is chosen to be the new bound. After this, τ_3 can be assigned to processor 1, and then τ_4 can be assigned to processor $\mathcal{M}_2$.

In practice, operating the same workload at a higher frequency may consume less energy due to the existence of idle power. Consider two consecutive frequencies, f' and f'', $f' < f''$, from the frequency set. If working at f' will consume more energy than working at f'', f' is regarded as an inefficient frequency. The modified AMBFF algorithm deals with an inefficient frequency as follows: whenever a inefficient frequency will be used as the frequency bound, the subsequent higher frequency is chosen instead.

Considering application-specific power consumption, after AMBFF, a swapping step is applied, trying to assign tasks with higher power coefficients to the core with lower speed settings.

Besides, AMBFF is applicable to both EDF and RM scheduling by using the schedulability test of EDF and RM, respectively, in the algorithm.

3.4 Sporadic Tasks

This section focuses on energy-aware scheduling of sporadic tasks on homogeneous multiprocessor platforms. Different from previous discussions, both [21] and [22] consider global scheduling for sporadic tasks. Nelis et al. [21] provide an optimal static scheduling method and a Multiprocessor One Task Extension (MOTE) scheme to further reduce energy consumption. Nelis and Goossens [22] reclaim the slack between the tasks' worst case execution times and actual case execution times to reduce energy consumption further. Zhang et al. [17] provide energy-efficient scheduling based on an optimal scheduling algorithm for sporadic tasks.

Global Energy-aware Scheduling: Nelis et al. [21] consider power-aware scheduling of sporadic constrained-deadline real-time tasks on homogeneous multiprocessor platforms, where global scheduling is adopted instead of partition-based scheduling. The active power consumption is assumed to only include dynamic power, and thus to achieve optimal energy conservation is to use speed as slow as possible while meeting all of the deadlines. Two distinct algorithms are proposed.

The first one provides an offline speed determination scheme that provides an identical speed for each processor, and this speed will not change during runtime. Firstly, the authors provide the minimal speed when using pure Earliest Deadline First (EDF) scheduling based on [23]'s EDF schedulability analysis. Although EDF is proven to be optimal for single processor scheduling, it is shown to not be optimal for multiprocessor scheduling. Then, the authors provide a superior offline speed determination method based on the schedulability analysis of $EDF^{(k)}$ [24]. Assuming that the tasks' densities λ_is are ranked as $\lambda_1 \geq \lambda_2 \geq \cdots \geq \lambda_n$, $1 \leq k \leq m$, the $EDF^{(k)}$ adopts the following rules: (a) for all $i < k$, τ_i jobs are assigned the highest priority (ties can be broken arbitrarily); (b) for all $i \geq k$, τ_i jobs are assigned priorities according to EDF (ties are also broken arbitrarily). The minimal speed is denoted by s_{opt}.

The second algorithm of [21] provides a runtime adaptive speed scheduling for all of the processors, where processors' speeds can all be different and can change independently with time. The technique is termed Multiprocessor One Task

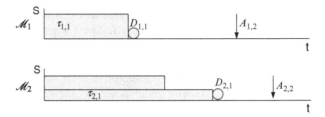

Fig. 3.19 MOTE scheme

Extension (MOTE). The idea is that the speed of a CPU can be reduced below s_{opt} during the execution of a job if the reduced speed doesn't change anything with respect to the schedule of the subsequent jobs that are scheduled on that CPU. Whenever a job is dispatched to a processor, it is calculated to what extent this job can be slowed down. Let t_{text} be the amount of time this processor will be required by another job (possibly a job from the same task). Then, it is safe to slow down the job such that it completes exactly at $min(t_c + D_i, t_{text})$, where t_c is the current time.

Consider the simple example in Fig. 3.19. $D_{1,1}$ and $D_{2,1}$ represent the deadlines of $\tau_{1,1}$ and $\tau_{2,1}$, respectively. $A_{1,2}$ and $A_{2,2}$ stand for the nearest arrival time of $\tau_{1,2}$ and $\tau_{2,2}$, respectively. When $\tau_{2,1}$ arrives, it has all the information on the other processors; based on this information, it can predicate when processor $\mathcal{M}_2$ will be requested by other jobs (possibly from the same task). In the example, t_{next} is calculated as $A_{2,2}$, which is the nearest possible arrival time of $\tau_{2,2}$. Thus, it should be slowed down to compete exactly at time $min(D_{2,1}, t_{next}) = D_{2,1}$, as shown in Fig. 3.19.

Global Energy-aware Scheduling with Slack Reclamation: For a similar problem, [22] provides an online slack reclamation scheme, termed MORA. The three main contributions of [22], compared to [21], are: [22] considers slack that results from the difference between the worst case execution requirement and the actual case execution requirement, non-ideal processors with discrete frequencies, and application-specific power consumptions. Denote $\tau_{i,j}$ as the jth job of task τ_i. Every $\tau_{i,j}$ is associated with two speeds, $s_{i,j}$ and $s_{i,j}^{off}$, where $s_{i,j}$ can change at any time during the system execution, and $s_{i,j}^{off}$ is the offline pre-computed execution speed of $\tau_{i,j}$ that ensures that all of the deadlines are met. MORA is based on reducing the execution speed $s_{i,j}$ of the jobs online, in order to save energy while still meeting all of the deadlines. MORA detects whenever the speed $s_{i,j}$ can be reduced by performing a comparison between the actual schedule and the offline schedule, which provides the $s_{i,j}^{off}$'s; MORA always refers to the offline schedule to produce the actual schedule. The mechanism that MORA reclaims slacks is that, when any job is finished in the actual schedule without consuming all of its WCET, the unused time should be used by starting the execution of any waiting job in that CPU earlier, and thus the execution speed of the selected waiting job can be reduced while still meeting all of the deadlines and not influencing the subsequent job executions.

Fig. 3.20 Offline VS MORA
scheduling. (a) Offline
scheduling. (b) MORA
scheme

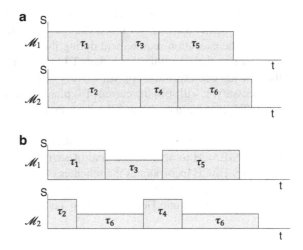

For example, consider the following six tasks, which are released at the same time, 0. $\tau_1 = (4, 40, 40)$, $\tau_2 = (5, 41, 41)$, $\tau_3 = (2, 42, 42)$, $\tau_4 = (2, 43, 43)$, $\tau_5 = (4, 44, 44)$, $\tau_6 = (4, 45, 45)$. According to EDF scheduling, from τ_1 to τ_6, tasks have descending priority. One valid offline schedule is provided. At time 0, the waiting jobs that will be assigned to processor $\mathcal{M}_1$ are τ_3 and τ_5, and those that will be assigned to $\mathcal{M}_2$ are τ_4 and τ_6. However, in the practical case, $\tau_{1,1}$ and $\tau_{2,1}$ complete without consuming their WCETs. Thus, after competing $\tau_{1,1}$, the scheduler should select a waiting job from τ_3 and τ_5 to be executed on $\mathcal{M}_1$. Thus, the selected task will be slowed down and will save energy. The selection from τ_3 and τ_5 is done by comparing the energy reductions of all available selections. The selection with the most energy saving is chosen. In the example shown in Fig. 3.20, we assume that τ_3 and τ_6 are selected, respectively.

Global Energy-aware Scheduling Based on an Optimal Scheduling for Sporadic Tasks: In [17], the authors propose an energy-efficient real-time scheduling algorithm for sporadic tasks, named LRE-DVFS-EACH. It is based on LRE-TL [25], which is an optimal scheduling algorithm for sporadic tasks. LRE-TL is an extension based on LLREF, and like LLREF, it uses the concepts of the T-L plane. One improvement of LRE-TL against LLREF is that it reduces the scheduling overhead of LLREF by eliminating some unnecessary migrations and preemptions. Another important contribution of LRE-TL is that it provides scheme that supports sporadic tasks. For periodic tasks, one task's next instance/job arrives exactly at the deadline of its current instance; however, for sporadic tasks, this is not the case. LRE-TL proposes a scheme to deal with the unpredicted instance/job arrivals of sporadic tasks. Initially, LRE-TL does the same thing as that of LLREF, namely, constructing the T-L plane according to task deadlines. For periodic tasks, there will be no new job arrivals during any T-L plane. However, there might be a new job arrival during a T-L plane for sporadic tasks (the *job arriving event* in a T-L plane is called event A). If the new job has a deadline that is beyond the current T-L plane, there is not much change compared to LLREF; LRE-TL just calculates the execution requirement during the current T-L plane. If the new job has a deadline that is within the current T-L plane, the LRE-TL will either check if the new job can execute non-preemptively without causing any deadline misses, or, if necessary, it will split the current T-L plane at the new job's deadline, forming two sub-T-L planes. In each sub-T-L plane, tasks are scheduled in a way similar to that of LLREF.

Based on the optimal LRE-TL algorithm, the authors in [17] proposes the LRE-DVFS-EACH to achieve the goal of saving energy. For feasible scheduling, LRE-DVFS-EACH incorporates the schemes of LRE-TL for handling events A, B, and C; for saving energy, it adopts a similar strategy to that in Independent SVFS, which classifies tasks into Heavy and Light Tasks.

Chapter 4
Scheduling on Heterogeneous DVFS Multiprocessor Platforms

Abstract As can be seen, a lot of research has been done for homogeneous platforms; comparatively, less has been done for heterogeneous platforms. As heterogeneous platforms are becoming more and more popular, energy-aware scheduling on heterogeneous platforms also needs further research focus. This chapter surveys existing works for energy-aware scheduling on heterogeneous platforms. It consists of three sections: frame-based tasks (Sun W, Sugawara T (2011) Heuristics and evaluations of energy-aware task mapping on heterogeneous multiprocessors. In: Proceedings of IEEE international symposium on parallel and distributed processing workshops and Phd forum, Alaska, May 2011, pp 599–607; Li D, Wu J (2012) Energy-aware scheduling for fame-based tasks on heterogeneous multiprocessor platforms. In: Proceedings of international conference on parallel processing, September 2012), tasks with precedence constraints (Lee YC, Zomaya AY (2009) Minimizing energy consumption for precedence-constrained applications using dynamic voltage scaling. In: Proceedings of the 9th IEEE/ACM international symposium on cluster computing and the grid, Shanghai, May 2009, pp 92–99), and periodic tasks (Hung C-M, Chen J-J, Kuo T-W (2006) Energy-efficient real-time task scheduling for a dvs system with a non-dvs processing element. In: Proceedings of the 27th IEEE international real-time systems symposium, Rio de Janerio, December 2006, pp 303–312; Yang C-Y, Chen J-J, Kuo T-W, Thiele L (2009) An approximation scheme for energy-efficient scheduling of real-time tasks in heterogeneous multiprocessor systems. In: Proceedings of design, automation test in Europe conference and exhibition, Nice, April 2009, pp 694–699; Chen J-J, Thiele L (2009) Task partitioning and platform synthesis for energy efficiency. In: Proceedings of the 15th IEEE international conference on embedded and real-time computing systems and applications, Beijing, pp 393–402; Chen J-J, Kuo T-W (2006) Allocation cost minimization for periodic hard real-time tasks in energy-constrained dvs systems. In: Proceedings of the 2006 IEEE/ACM international conference on computer-aided design, San Jose, pp 255–260). Actually, both (Yang C-Y, Chen J-J, Kuo T-W, Thiele L (2009) An approximation scheme for

D. Li and J. Wu, *Energy-aware Scheduling on Multiprocessor Platforms*,
SpringerBriefs in Computer Science, DOI 10.1007/978-1-4614-5224-9_4,
© The Author(s) 2013

energy-efficient scheduling of real-time tasks in heterogeneous multiprocessor systems. In: Proceedings of design, automation test in Europe conference and exhibition, Nice, April 2009, pp 694–699) and (Chen J-J, Thiele L (2009) Task partitioning and platform synthesis for energy efficiency. In: Proceedings of the 15th IEEE international conference on embedded and real-time computing systems and applications, Beijing, pp 393–402) consider frame-based tasks and periodic tasks simultaneously; we put them under the category of periodic tasks. To the best of our knowledge, little has been done for energy-aware sporadic task scheduling on heterogeneous multiprocessor platforms.

4.1 Frame-Based Tasks

Heuristics for Minimizing Energy Consumption or Considering Makespan and Energy Consumption Simultaneously: In [26], the authors address the problem of mapping a set of frame-based tasks to heterogeneous multiprocessors. The heterogeneous DVFS processors are assumed to have the same discrete voltage levels. The maximal voltage is v_1, and the voltage at level k is v_k; there are K different voltage levels. By assigning task set $\mathscr{T} = \{\tau_1, \tau_2, \cdots, \tau_n\}$ to the m processors, a task mapping is created and denoted by $\mathscr{P}$, which consists of n tuples $\{i \in [1, n], j \in [1, m], k \in [1, K]\}$. Denote $\text{ETC}^3 = \{t_{i,j,k}\}^{n \times m \times K}$ as the Expected Time to Compute matrix, and $\text{EEC}^3 = \{e_{i,j,k}\}^{n \times m \times K}$ as the Expected Energy to Complete matrix. $t_{i,j,k}$ and $e_{i,j,k}$ represent the time and energy required to complete task τ_i on processor $\mathscr{M}_j$ at voltage level v_k, respectively. Let $t_{i,j,1}$ be the execution time and $e_{i,j,1}$ be the energy consumption at full processor speed; according to their simple assumptions, $t_{i,j,k} = \frac{v_1}{v_k} t_{i,j,1}$, $e_{i,j,k} = (\frac{v_k}{v_1})^2 e_{i,j,1}$. The schedule length of a set of tasks under a given partition is denoted by:

$$D = max_{j \in [1, m]} \left(\sum_{i \in [1, n], k \in [1, K]} t_{i,j,k}[\{i, j, k\} \in \mathscr{P}] \right).$$

which is the greatest timespan among all of the processors. The total energy consumption is denoted by:

$$E = \sum_{\{i, j, k\} \in \mathscr{P}} e_{i,j,k}.$$

Two approaches are proposed for the scheduling problem:

A1: First, find the minimum schedule length and a feasible mapping/partition if all processors operate at the maximal voltage level v_1 by the Min-min heuristic; then, adjust the supply voltage for each mapped task to appropriately extend the schedule length and to achieve the minimal energy consumption.

A2: Simultaneously consider schedule length and total energy consumption in each step.

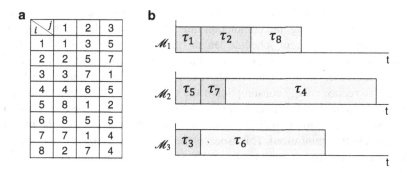

Fig. 4.1 An example for Min-min heuristic. (**a**) $t_{i,j,1}$. (**b**) Partition by Min-min heuristic

Here, we only present four heuristics that are proposed in the paper: H1 and H2 for A1, H3 and H4 for A2.

The first step of H1 and H2 is the same: find a mapping by the Min-min heuristic at the maximal voltage level. The Min-min heuristic adopts the strategy of selecting a task to assign to a processor, if the resulting completion time on this processor is the minimal among all processors (also after assigning a task). The following example in Fig. 4.1 shows how the Min-min heuristic works. Let $t_{i,j,1}$ be the time to complete task τ_i on processor $\mathcal{M}_j$ at full speed. All of the $t_{i,j,1}$ values for eight tasks on three processors are given in Fig. 4.1a. Consider the time when τ_1, τ_2, τ_3, τ_5, and τ_7 have been assigned. Denote $PATH_j$ as the current accumulative execution time on processor $\mathcal{M}_j$. We have $PATH_1 = 3$, $PATH_2 = 2$, $PATH_3 = 1$; the remaining tasks are τ_4, τ_6, τ_8. After some simple calculations, we have $min(PATH_1 + t_{i,1,1}) = 3 + t_{8,1,1} = 5$, $min(PATH_2 + t_{i,2,1}) = 2 + t_{6,2,1} = 7$, and $min(PATH_3 + t_{i,3,1}) = 1 + t_{6,3,1} = 6$; obviously, the minimal among these three is $min(PATH_1 + t_{8,1,1}) = 5$, so in the next step, task τ_8 is assigned to processor $\mathcal{M}_1$. The final assignment, according to the Min-min heuristic, is shown in Fig. 4.1b.

The second step of H1 and H2 is to optimize the supplied voltage for each task without violating the predefined schedule length. For non-ideal DVFS processors, finding the optimal voltage setting for each task on a processor $\mathcal{M}_j$ is an Integer Linear Programming (ILP) problem. Finding a solution by H1 is equivalent to searching a full K-ary tree with the height equal to the number of tasks on a processor. H2 is proposed to simplify H1. After mapping a task, the completion time of each processor t_j^e and the schedule length D at the maximal voltage level are known and a reasonable schedule length constraint $D_c \geq D$ is decided. If there exists voltage level $v_1 \frac{t_j^e}{D_c}$, the schedule length will be exactly D_c if executing at this voltage level. There must be $v_l, v_h \in \{v_1, v_2, \cdots, v_K\}$, such that $v_l \leq v_1 \frac{t_j^e}{L_c} \leq v_h$. Each task in $\mathcal{M}_j$ only needs to choose v_l or v_h and the schedule length constraint will be guaranteed. Thus, the K-ary tree search is reduced to a binary tree search, though the height of the tree is still the number of tasks on the processor.

H3 first searches for the next task to map like the Min-min heuristic. If, according to the Min-min mapping, the completion time of this task is less than a threshold which extends the current schedule length (namely, $\alpha D_c, \alpha > 1$). The supplied voltage to this task will be decreased and meanwhile the completion time of this task will increase, but not by too much. It can be seen that the intuition of H3 lies in trying to reduce energy consumption if the schedule length is not increasing too fast. On the other hand, it is also applicable to map each next task in a way that its energy consumption is with a given threshhold and increment of the resulting schedule length is minimized. This aspect is the idea of H4. If such an assignment is infeasible, the task at hand will be mapped to a processor with minimum completion time.

A Novel Approach for Minimizing Energy Consumption: In our previous work [27], we address the problem of scheduling a set of frame-based tasks on heterogeneous multiprocessor platforms with the goal of minimizing the overall energy consumption while still meeting all tasks' deadlines. We propose a Relaxation-based Iterative Rounding Algorithm (RIRA) for the problem. This work falls into the category of partition-based scheduling, and after partitioning, task migration and preemption are not allowed. For the heterogeneity of the platform and difference among tasks, an execution efficiency matrix $\lambda_{n \times m}$ is defined, where $\lambda_{i,j}$ represents the execution efficiency of processor j when it is used to execute task τ_i. In other words, for a task τ_i with worst case execution cycles $WCEC_i$, the time required to finish the task on processor j at frequency f can be calculated as $WCEC_i/(\lambda_{i,j}f)$. Actually, three types of heterogeneous platforms are considered, namely, dependent platforms without runtime adjusting, dependent platforms with runtime adjusting, and independent platforms. For simplicity we will only explain the algorithm on the first type of platform in detail.

We first consider the optimal frequency setting if we have already had a task partition. Let binary variables $x_{i,j}$ be 1 if task τ_i is assigned to processor $\mathcal{M}_j$, and 0 otherwise. A given partition can be represented by a binary matrix $x_{n \times m}$. We denote the shared frequency among all of the processors during the whole time by f. Then, the time when processor $\mathcal{M}_j$ will complete its workload can be calculated as $\frac{1}{f} \sum_{i=1}^{n} \frac{x_{i,j}WCEC_i}{\lambda_{i,j}}$. The shared frequency should guarantee that all processors will finish the tasks assigned to it before the deadline:

$$\frac{1}{f} \sum_{i=1}^{n} \frac{x_{i,j}WCEC_i}{\lambda_{i,j}} \leq D, \forall j = 1, 2, \cdots, m.$$

The energy consumption on the jth processor $\mathcal{M}_j$ can be calculated as:

$$E_j = f^3 \left(\frac{1}{f} \sum_{i=1}^{n} \frac{x_{i,j}WCEC_i}{\lambda_{i,j}} \right) = f^2 \sum_{i=1}^{n} \frac{x_{i,j}WCEC_i}{\lambda_{i,j}}$$

Thus, to achieve a partition with the goal of saving energy, the problem can be formulated as the following binary integer programming problem:

$$\min \quad E_{total} = f^2 \sum_{j=1}^{m} \left(\sum_{i=1}^{n} \frac{x_{i,j} WCEC_i}{\lambda_{i,j}} \right)$$

$$s.t. \quad \sum_{i=1}^{n} \frac{x_{i,j} WCEC_i}{\lambda_{i,j}} - fD \leq 0, \forall j = 1, 2, \cdots, m.$$

$$x_{i,j} = 0 \ or \ 1, \forall i = 1, 2, \cdots, n; j = 1, 2, \cdots, m.$$

$$\sum_{j=1}^{m} x_{i,j} = 1, \forall i = 1, 2, \cdots, n.$$

$$f \geq 0.$$

where the optimization variables are the shared frequency f and the binary matrix $x_{n \times m}$. It is known that binary integer programming problems are NP-complete. Thus, we consider relaxing the binary variables $x_{i,j}$'s to be any fraction in $[0, 1]$. Denote the relaxed optimization problem by P_1, which is a convex optimization problem that can be solved by the well-known interior point method in polynomial time (in terms of the input problem size under a given precision requirement) [28]. The optimization variables of P_1 are the shared frequency f and the relaxed assignment matrix $x_{n \times m}$. Here, $x_{i,j}$ represents the percentage of task τ_i that should be assigned to processor M_j to achieve the minimal overall energy consumption.

Our intuition is that if we assign tasks in a way that is "closest" to the optimal solution (for the relaxed problem), we will achieve a better partition in terms of overall energy consumption. Introduce a reference execution time matrix, $t_{n \times m}$, where $t_{i,j} = WCEC_i / \lambda_{i,j}$. Further define task τ_i's average execution requirement as $AER_i = \frac{\sum_{j=1}^{m} t_{i,j}}{m}$. Without loss of generality, assume that all of the tasks are sorted in descending order of their average execution requirement. This is also the order that we will assign tasks in. Our intuition is that the task with the greatest average execution requirement is the most influential task in terms of both schedulability and energy consumption, and should be assigned first.

The solutions $x_{1,1}, x_{1,2}, \cdots, x_{1,n}$ for P_1 indicate the optimal assignment of the most influential task τ_1. Then, we find the maximum among $x_{1,1}, x_{1,2}, \cdots, x_{1,n}$, denoted by x_{1,j^*}, and assign τ_1 to processor M_{j^*}. Denote the final assignment matrix for the task set by $Assign_{n \times m}$. Then, we have $Assign_{1,j} = 0, \forall j \neq j^*$ and $Assign_{1,j^*} = 1$.

Before assigning the next most influential task τ_2, we need to update the optimization problem first. Notice that we have already assigned task τ_1 to processor M_{j^*}, which means that $x_{1,j} = 0, \forall j \neq j^*$ and $x_{1,j^*} = 1$. The updated optimization problem can be formulated as:

$$\min \quad E_{total} = f^2 \sum_{i=1}^{m} \left(\sum_{i=1}^{n} \frac{x_{i,j} WCEC_i}{\lambda_{i,j}} \right)$$

$$s.t. \quad \sum_{i=1}^{n} \frac{x_{i,j} WCEC_i}{\lambda_{i,j}} - fD \leq 0, \forall j = 1, 2, \cdots, m.$$

Table 4.1 Example for RIRA

i	$WCEC_i$	$\lambda_{i,j}$			$t_{i,j}$		
		$j=1$	$j=2$	$j=3$	$j=1$	$j=2$	$j=3$
1	7	0.7	0.4	0.1	10	17.5	70
2	8	0.5	0.2	0.3	16	40	26.67
3	3	0.4	0.1	0.2	7.5	30	15
4	5	0.5	0.2	0.4	10	25	12.5
5	9	0.6	0.9	0.7	15	10	12.86
6	5	0.8	0.3	0.5	6.25	16.67	10
7	4	0.3	0.9	0.6	13.33	4.44	6.67
8	4	0.4	0.6	0.8	10	6.67	5

$$0 \le x_{i,j} \le 1, \forall i = 2, \cdots, n; \forall j = 1, 2, \cdots, m.$$

$$\sum_{j=1}^{m} x_{i,j} = 1, \forall i = 2, \cdots, n$$

$$f \ge 0.$$

Denote this optimization problem by P_2 since it will provide the solution for assigning task τ_2. Notice that P_2 is quite different from P_1 because now, $x_{1,1}, x_{1,2}, \cdots, x_{1,m}$ have fixed values, namely, $x_{1,j} = 0, \forall j \ne j^*$, and $x_{1,j^*} = 1$. The optimization variables in P_2 only include the optimal frequency f and $x_{2,1}, x_{2,2}, \cdots, x_{2,m}, x_{3,1}, x_{3,2}, \cdots, x_{3,m}, \cdots, x_{n,1}, x_{n,2}, \cdots, x_{n,m}$. Thus, we can assign τ_2 according to $x_{2,1}, x_{2,2}, \cdots, x_{2,m}$ (solved for P_2) based on the maximum value among them, which is similar to what we do to assign task τ_1. Then, we can update the optimization problem as P_3, keeping in mind that we have already assigned task τ_1 and τ_2; solve it and assign task τ_3. Then, solve P_4 to assign $\tau_4; \cdots$; solve P_i to assign $\tau_i; \cdots$. Repeat the process until we finish assigning $(n-1)$ tasks. It can be seen that, in some cases, assigning the last task according to this iterative scheme may not be optimal. Thus, for the last task, the assignment, which can achieve the minimal overall energy consumption among all possible assignments for the last task, is selected.

An illustrative example is provided below, which deals with assigning eight tasks to three processors. Tasks' $WCEC$'s and the processor efficiency matrix $\lambda_{8 \times 3}$ are given in Table 4.1. A reference execution time matrix is denoted by $t_{8 \times 3}$, where $t_{i,j} = WCEC_i / \lambda_{i,j}$, which is also provided in the same table.

The proposed RIRA first solves the original optimization problem P_1; assign τ_1 according to solutions $x_{1,1}, x_{1,2}, x_{1,3}$ (solved for P_1). Then, it updates the optimization problem to P_2, solves it, and assigns τ_2 according to solutions $x_{2,1}, x_{2,2}, x_{2,3}$ (solved for P_2). Repeat the above process until it assigns seven tasks; for the last task, it selects the assignment that achieves the minimal energy consumption. Relevant solutions are shown in Table 4.2. Figure 4.2 shows the partition by the proposed RIRA.

Table 4.2 Iterative assigning by RIRA

	Relaxed Assignment $x_{i,j}$ solved for P_i			$Assign_{8 \times 3}$		
i	$j = 1$	$j = 2$	$j = 3$	$j = 1$	$j = 2$	$j = 3$
1	0.2920	0.7080	0	0	1	0
2	1	0	0	1	0	0
3	0.99984	0.00001	0.00015	1	0	0
4	0.00013	0.00001	0.99986	0	0	1
5	0	0.5379	0.4621	0	1	0
6	0.6504	0	0.3496	1	0	0
7	0	0.5062	04938	0	1	0
8	–	–	–	0	0	1

Fig. 4.2 Final partition by RIRA

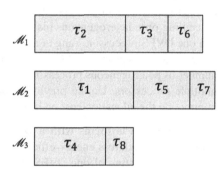

After a partition, the processors are slowed down dependently such that the processor with the greatest completion time meets the predefined deadline D exactly.

4.2   Tasks with Precedence Constraints

Scheduling Simultaneously Considering Makespan and Energy Consumption: Lee and Zomaya [29] considers scheduling precedence-constrained tasks/applications on heterogeneous DVFS multiprocessor platforms. Each processor has a distinct set of voltage levels. Power consumption is regarded as being proportional to the square of the supply voltage. Besides, when consecutive tasks are assigned to different processors, any communication between them is not neglected. Applications/tasks under consideration are not deadline-constrained. Thus, the scheduling quality for the task set is measured explicitly, considering both makespan and energy consumption. The authors devise an Energy-Conscious Scheduling (ECS) heuristic with a newly defined objective function: *relative superiority*, which takes both the makespan and energy consumption into account. The tasks are firstly sorted in an order that satisfies the original precedence constraints. Then, tasks are considered according to this order, one by one, to be assigned to a processor with the best *relative superiority* value.

After this procedure, a Makespan Conservative Energy Reduction (MCER) technique is incorporated. For each task, MCER considers all of the other combinations of (host processor, voltage supply level) to check whether any one of these combinations reduces the energy consumption of the task without increasing the current makespan. If such a combination exists, the task will switch to use this combination.

4.3 Periodic Tasks

This section surveys existing works on energy-aware scheduling of periodic tasks on heterogeneous multiprocessor platforms. Hung et al. [30] consider energy-aware scheduling on a heterogeneous platform with one non-DVFS Processor Unit (PU) and one DVFS processor. Yang et al. [31] deal with platforms with a fixed number of heterogeneous processors. Chen and Thiele [32] consider platforms with a fixed number of heterogeneous processor types, while one processor type may still have multiple processors. Unlike previous discussions, [33] addresses the problem of minimizing the allocation cost on heterogeneous platforms.

Scheduling on Platforms with a DVFS processor and a non-DVFS PU: The authors in [30] address energy-efficient periodic task scheduling on heterogeneous platforms, consisting of one DVFS processor and one non-DVFS Processing Unit (PU). The non-DVFS PU can be either workload-dependent, where its energy consumption in the hyper-period H is $P_2 U_2 H$, or workload-independent, where its energy consumption in the hyper-period H is $P_2 H$. P_2 is the constant power, and U_2 is the utilization assigned to the non-DVFS PU. Denote μ_i as the utilization of task τ_i on the non-DVFS PU.

For ideal DVFS processors, when the non-DVFS PU is workload-independent, the optimization problem can be formulated as follows:

$$minimize \qquad \sum_{\tau_i \in \mathscr{T}} x_i C_i / T_i$$
$$s.t. \ \sum_{\tau_i \in \mathscr{T}} \mu_i x_i \geq \left(\sum_{\tau_i \in \mathscr{T}} \mu_i \right) - 1$$

where, x_i is 1 if τ_i is assigned to the DVFS processor and is 0 if τ_i is assigned to the non-DVFS PU. Our intuition here is that, if a task has high computation demand on the DVFS processor, but a low utilization on the non-DVFS PU, it will be a good candidate to be assigned to the non-DVFS PU. According to this intuition, Algorithm GREEDY is proposed and works as follows: (1) sort tasks in a non-decreasing order of $\frac{\mu_i}{C_i/T_i}$ and initialize the task set assigned to the non-DVFS PU, $\mathscr{T}_2$ as empty; (2) for $i = 1, \cdots, n$, if $\mu_i + \sum_{\tau_j \in \mathscr{T}_2} \mu_j \leq 1$, τ_i is added to $\mathscr{T}_2$; (3) assign all of the tasks in $\mathscr{T}_2$ to the non-DVFS PU and all of the remaining tasks to the

Fig. 4.3 Algorithm
GREEDY

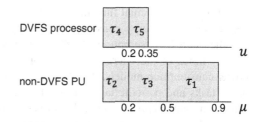

DVFS processor. Algorithm GREEDY is reasonable but does not provide a worst case guarantee. Thus, based on Algorithm GREEDY, another enhanced algorithm, E-GREEDY, is proposed and is proven to be an approximation algorithm.

Consider the following five tasks: $\tau_1 = (6,10,10)$, $\tau_2 = (4,8,8)$, $\tau_3 = (12,20,20)$, $\tau_4 = (2,10,10)$, $\tau_5 = (1.8,12,12)$; $\mu_1 = 0.4$, $\mu_2 = 0.2$, $\mu_3 = 0.3$, $\mu_4 = 0.15$, $\mu_5 = 0.2$. Thus, their $\frac{\mu_i}{C_i/T_i}$ values are: 0.67, 0.4, 0.5, 0.75, 1.33. Algorithm GREEDY works as following: sort the tasks in non-decreasing order of $\frac{\mu_i}{C_i/T_i}$: $\tau_2, \tau_3, \tau_1, \tau_4, \tau_5$. Thus, τ_2, τ_3, τ_1 should be assigned to the non-DVFS PU, and the utilization on the non-DVFS PU will be 0.9; since $\mu_4 + 0.9 = 1.05 > 1$, no more tasks will be assigned to the non-DVFS PU. τ_4 and τ_5 will be assigned to the DVFS processor. The final assignment is shown in Fig. 4.3.

When the non-DVFS PU is workload-dependent, the reduction of energy consumption on the DVFS processor and the increase of the energy consumption on the non-DVFS PU must be evaluated, while considering the assignment of a task to the non-DVFS PU. Another algorithm, S-GREEDY, is proposed. Initially, tasks are still sorted in a non-decreasing order of $\frac{\mu_i}{C_i/T_i}$. Put all of the tasks on the DVFS PE as the initial solution. According to the sorted order, consider the assignment of task τ_i in the ith iteration. According to the solution so far, if moving more portions of task τ_i to the non-DVFS PU can reduce the energy consumption further, then assign task τ_i to the non-DVFS PE; otherwise, assign τ_i on the DVFS processor. After n iterations, a task assignment for task set $\mathscr{T}$ can be achieved.

Scheduling on Platforms with a Fixed Number of Heterogeneous Processors:
In [31], the authors consider how to partition real-time tasks on a heterogeneous platform to achieve the minimal energy consumption, where, after partitioning, task migration is not allowed. Let $E_j(U_j)$ be the energy consumed on $\mathscr{M}_j$ with workload U_j. Assuming that the energy consumption of a given processor with a higher workload is larger than that with a lower workload, they propose an approximation scheme for different power models and task types when the number of processors is a constant.

The authors present an important inequality that many practical systems satisfy:

$$if \qquad 0 < U_j, (1+\delta)U_j \le 1, 0 \le \sigma \le \delta$$
$$then\ E_j((1+\sigma)U_j) \le E_j((1+\delta)U_j) \le (1+\varepsilon)E_j(U_j)$$

where δ is a polynomial function of constant ε in a specific range. The inequality here means that energy consumption on $\mathcal{M}_j$ is a non-decreasing function of U_j, and the energy consumption of $\mathcal{M}_j$, with workload $(1 + \delta)U_j$, is no more than $(1 + \varepsilon)$ times of the energy consumption with workload U_j. Based on various power consumption models, it is shown that this inequality is true for both frame-based tasks and periodic tasks on ideal DVFS shutdown-disabled processors and shutdown-enabled processors without overhead ($E_j^{ov} = 0$); it is also true for frame-based tasks on ideal DVFS shutdown-enabled processors with overhead ($E_j^{ov} \neq 0$), but it is not true for periodic tasks.

A dynamic programming approach is applied to tackle the problem, and it is shown to be NP-hard. The basic idea of the dynamic programming is to keep tracing a set of states S, where each state stands for a partition $\mathcal{P}$ for a subset of task set $\mathcal{T}$. A state that stands for a partition $\mathcal{P}$ is presented by a tuple $p = (U_1^P, U_2^P, \cdots, U_m^P)$. The initial set of states S_0 consists of one state $p = (0, 0, \cdots, 0)$. When τ_i is considered, a set S_i of new states will be generated according to the existing states S_{i-1}. For each state p in S_{i-1}, the dynamic programming constructs m new states $(U_1^P + u_{i,1}, U_2^P, \cdots, U_m^P)$, $(U_1^P, U_2^P + u_{i,2}, \cdots, U_m^P)$, $\cdots$, $(U_1^P, U_2^P, \cdots, U_m^P + u_{i,m})$ and adds them into set S_i. Eventually, all feasible partitions are in the set S_n. Then, based on the energy inequality, which holds for various systems, a pruning approach is applied to reduce considerable states in every step to achieve $S_1, S_2, \cdots, S_n$. The pruning approach modifies the dynamic programming process to be in polynomial time and limits the sacrifice in the optimality of the final result. At the end, it is proven that the algorithm, termed Algorithm MTRIM, has an approximation ratio of $(1 + \varepsilon)$ if the derived solution is feasible, where ε is a parameter adopted in the procedure.

Scheduling on Platforms with a Fixed Number of Heterogeneous Processor Types: The authors in [32] also explore how to partition tasks and select processors on heterogeneous platforms with the goal of saving energy, where online migration is not allowed. The platform under consideration has m processor types, and each type may still have multiple processors. Two kinds of processor types are considered: processors without the capability of DVFS, whose power consumption is a constant, and processors with the capability of DVFS. The problem of energy-efficient task partitioning with heterogeneous processor types is termed as the ETHE problem when the number of processors of type $\mathcal{M}_j$ is not restricted; when the number of available processors of type $\mathcal{M}_j$ is upper bounded by N_j^{up}, it is termed as the R-ETHE problem.

Let $E_j(\mathcal{T}_j)$ be the minimal total energy consumption in hyper-period H for scheduling a subset $\mathcal{T}_j$ of task set $\mathcal{T}$, while meeting all of the deadlines, on a processor of type $\mathcal{M}_j$. Much like in [31], the authors in this paper first present a statement that various systems satisfy:

$$if \ \textstyle\sum_{\tau_i \in \mathcal{T}_{j,a}} u_{i,j} \leq \sum_{\tau_i \in \mathcal{T}_{j,b}} u_{i,j}$$

$$then \quad E_j(\mathcal{T}_{j,a}) \leq E_j(\mathcal{T}_{j,b})$$

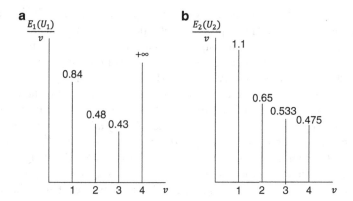

Fig. 4.4 All possible values of $E(U)/v$

Define the average energy consumption of tasks on a processor as the energy consumption of the processor divided by the number of tasks executed on this processor. A Minimum Average Energy First (MAEF) strategy is adopted to tackle the task assignment and processor allocation problem, with the intuition that to allocate a processor that can schedule more tasks with less energy consumption is preferred. Let $\mathcal{T}^{rest}$ be the set of tasks that are not assigned to allocated processors yet, where $\mathcal{T}^{rest}$ is initialized as $\mathcal{T}$. Because energy consumption is an increasing function of the total utilization of the tasks assigned to the processor, to assign v tasks in $\mathcal{T}^{rest}$ to a processor of type $\mathcal{M}_j$ with the minimum average energy consumption, it should choose those v lower utilization tasks in $\mathcal{T}^{rest}$ on a processor of type $\mathcal{M}_j$.

Let $\mathcal{T}^{rest}_{v,j}$ be the set of tasks (in $\mathcal{T}^{rest}$) that contains the v lower utilization tasks when assigned to a processor of type $\mathcal{M}_j$. By considering all of the possible assignments for $j = 1, 2, \cdots, m$ and $v = 1, 2, \cdots, |\mathcal{T}^{rest}|$, the processor allocation and task assignment with the minimal average energy consumption $\mathcal{T}^{rest}_{v^*,j^*}/v^*$ can be found. Thus, a processor of type $\mathcal{M}_{j^*}$ is allocated to $\mathcal{T}^{rest}_{v^*,j^*}$. Then, $\mathcal{T}^{rest}$ is updated by subtracting $\mathcal{T}^{rest}_{v^*,j^*}$ from $\mathcal{T}^{rest}$. The above procedure is repeated until all of the tasks are assigned to an allocated processor.

Consider the following example with four tasks whose utilizations are: $u_{1,1} = 0.3$, $u_{2,1} = 0.2$, $u_{3,1} = 0.2$, $u_{4,1} = 0.75$, $u_{1,2} = 0.1$, $u_{2,2} = 0.2$, $u_{3,2} = 0.3$, $u_{4,2} = 0.3$. Assume that $E_1(U_1) = U_1^2 + 0.8$, $E_2(U_2) = U_2 + 1$.

The MAEF strategy works as follows: for $j = 1$, sort tasks according to their $u_{i,1}$ values in ascending order: $u_{2,1} = 0.2$, $u_{3,1} = 0.2$, $u_{1,1} = 0.3$, $u_{4,1} = 0.75$. Choose the former v lower utilization tasks and calculate $E_1(U_1)/v$ for $v = 1, 2, 3, 4$: 0.84, 0.48, 0.43, $+\infty$ (since the accumulated utilization is $1.25 \geq 1$); for $j = 2$, sort tasks according to their $u_{i,2}$ values in ascending order: $u_{1,2} = 0.1$, $u_{2,2} = 0.2$, $u_{3,2} = 0.3$, $u_{4,2} = 0.3$. Choose the former v lower utilization tasks and calculate $E_2(U_2)/v$ for $v = 1, 2, 3, 4$: 1.1, 0.65, 0.533, 0.475. To show it clearly, all E_j/v ($j = 1, 2, v = 1, 2, 3, 4$) values are given in Fig. 4.4. Obviously, $E_1(U_1)/3$ is the minimal. Thus, τ_2, τ_3, τ_1 are assigned to a processor of type $\mathcal{M}_1$. $\mathcal{T}^{rest}$ is updated as $\{\tau_4\}$. Then, assigning τ_4 to a processor of type $\mathcal{M}_2$ is better than assigning to a processor of type $\mathcal{M}_1$. The final assignment is as follows: τ_1, τ_2, τ_3 are assigned to a processor of type $\mathcal{M}_1$, τ_4 is assigned to a processor of type $\mathcal{M}_2$.

Then, it is proven that the Algorithm MAEF has an approximation ratio of $(1 + \ln n)$ for the ETHE problem, where n is still the number of tasks. For the R-ETHE problem, Algorithm MAEF is modified as follows: when determining the minimum average energy consumption, if it has already allocated N_j^{up} (which is the maximum number of processors of type $\mathcal{M}_j$) processors of type $\mathcal{M}_j$, it just ignores the possibility of allocating one more processor of type $\mathcal{M}_j$. If the algorithm fails to provide a feasible solution, the algorithm in [34] is applied to try to achieve a feasible solution, because, at this time, feasibility matters more than saving energy.

Allocation Cost Minimization: For periodic tasks on heterogeneous platforms, the objective of [33] is to allocate processors (of different processor types) with the minimal allocation cost under timing and energy consumption constraints. $\mathcal{M} = \{\mathcal{M}_1 \cdots, \mathcal{M}_m\}$ is the set of all processor types. Each processor type has multiple processors. Allocating η processors of type $\mathcal{M}_j$ requires a cost of $\eta \mathscr{C}_j$. The available speeds and the power consumption function $P_j()$ of processor type $\mathcal{M}_j$ are specified. Given that the maximum energy consumption of task set $\mathcal{T}$ in the hyper-period H is bounded by E^{max}, the multiprocessor allocation for energy-constrained real-time scheduling (MARTS) problem is proven to be NP-hard in a strong sense, even when there is only one processor type for either ideal or non-ideal processors.

For non-ideal processors, the MARTS problem can be formulated as an integer linear programming problem. Then, a series of relaxation techniques are applied to achieve a feasible schedule and a proper allocation of processors in polynomial time. Specifically, by applying a parametric relaxation, a polynomial-time approximation algorithm is developed, based on a rounding technique. The algorithm to find all of the feasible solutions is termed as Algorithm ROUNDING, and another enhanced algorithm, E-ROUNDING, is developed to find the schedule with the minimum cost. For ideal processors, the available speeds are divided into a user-defined spectrum with a number of discrete speeds on each processor type. Then, Algorithm ROUNDING and E-ROUNDING are applied to allocate processors and assign tasks.

Chapter 5
Related Work

Abstract This chapter describes some works that are similar to, or very close to, what we do in this book. These include a brief survey on energy-aware scheduling on both uniprocessor and multiprocessor platforms, two research summaries of specific issues with multiprocessor energy-aware scheduling, an extensive energy-aware scheduling survey on uniprocessor systems that focuses on battery-powered devices, as well as a pure scheduling survey on multiprocessor platforms.

Energy-aware scheduling has been a hot research topic for a long time. The most recent comprehensive survey on energy-aware scheduling is Chen and Kuo [35]. It surveys research works on both uniprocessor and multiprocessor DVFS platforms. However, it focuses mainly on uniprocessor platforms, and only a small part is on multiprocessor platforms; overall, issues on multiprocessor platforms are not extensively covered.

Chen et al. [36] summarize previous works, which provide approximation algorithms for several issues on energy-efficient scheduling on multiprocessor DVFS platforms; [37] summarizes several existing works that consider leakage current and non-DVS components. However, only a limited number of issues are covered in both [36] and [37]. Also, they do not provide comprehensive classifications for energy-aware scheduling problems on multiprocessor platforms.

In [38], the authors survey previous studies on energy-efficient scheduling on uniprocessor systems. They focus on techniques proposed for battery-powered devices, aiming at using the right amount of energy in the right place at the right time. Some imperative concepts are also provided, such as periodic tasks, sporadic tasks, fixed/dynamic priority, preemptive and non-preemptive, etc. Their major contribution is that they extend their work to deal with the problem of scheduling tasks on platforms that are rechargeable [39]. They survey some studies that deal with uniprocessor rechargeable systems. Besides, throughout their work, they only deal with uniprocessor systems, though their works may share some light on multiprocessor platforms. As can be seen, their work is quite different from ours.

D. Li and J. Wu, *Energy-aware Scheduling on Multiprocessor Platforms*,
SpringerBriefs in Computer Science, DOI 10.1007/978-1-4614-5224-9_5,
© The Author(s) 2013

Davis and Burns [40] provides a survey of scheduling algorithms and schedulability analyses for multiprocessor systems. However, this work only covers scheduling algorithms for homogeneous multiprocessor platforms. Although it is not about energy-aware scheduling, we have noticed that most energy-aware scheduling algorithms are based on basic scheduling algorithms.

Our work in this book provides a comprehensive classification of energy-aware scheduling research on both homogeneous and heterogeneous multiprocessor platforms and presents clear definitions of several typical issues. Detailed techniques and methods that are used to solve various problems are also included.

Chapter 6
Conclusion and Future Directions

Abstract We conclude our book in this chapter. Based on our work above, a generalized description of energy-aware scheduling problems is presented. We also retell the overall hierarchy of our work. After that, we predict future directions on energy-aware scheduling on multiprocessor platforms according to three aspects, namely, platform models, task models, and other related concepts.

As we can see, energy-aware scheduling problems on multiprocessor platforms are intrinsically optimization problems, namely, minimizing energy consumption under a time constraint (performance constraint), minimizing execution time (maximizing performance) under an energy consumption constraint, especially for tasks with precedence constraints/framed based tasks, and minimizing the allocation cost under both energy consumption and time constraints. The formulations of these optimization problems greatly depend on the platform, task model, and various assumptions, so do the solutions for these problems. When models are simple, the optimization problems may have explicit analytical solutions, based on which, optimal scheduling can be derived directly. When models become complicated, the optimization problems also become complicated or even NP-hard. Under this situation, instead of finding the optimal solution, heuristic algorithms and approximation algorithms can be achieved based on the solutions of the formulated optimization problems or their relaxed versions.

As a brief conclusion, this book provides a comprehensive survey of what has been done regarding energy-aware scheduling on multiprocessor platforms. Two types of platforms are considered: homogeneous platforms and heterogeneous platforms. For each type of platform, different types of tasks are considered: frame-based tasks, tasks with precedence constraints, periodic tasks, and sporadic tasks. Under this classification, discussions on other issues, such as slack reclamation, fixed/dynamic priority scheduling, partition-based/global scheduling, and application-specific power consumption are also provided. For each specific problem, the state-of-the-art scheduling algorithms are provided in detail.

D. Li and J. Wu, *Energy-aware Scheduling on Multiprocessor Platforms*,
SpringerBriefs in Computer Science, DOI 10.1007/978-1-4614-5224-9_6,
© The Author(s) 2013

We believe that energy-aware scheduling on multiprocessor platforms will continue to be an important issue in modern computational systems, and thus deserves much research interest. We know that energy-aware scheduling on multiprocessor platforms, to a large extent, is based on basic scheduling algorithms and schedulability analysis on similar platforms. For partition-based scheduling, well understood scheduling algorithms for uniprocessors are widely used directly. For example, the Earliest Deadline First (EDF) schedulability condition and the Rate Monotonic (RM) schedulability condition [41]. For global scheduling, more complex scheduling algorithms and schedulability analyses are also introduced into energy-aware global scheduling. A typical example has been shown by Nelis et al. [21], Bertogna et al. [23], and Goossens et al. [24]. Thus, before considering energy-aware scheduling, we should first study basic scheduling algorithms and schedulability analyses under similar assumptions.

The future directions can be given based on the three aspects, namely, platform models, task models, and other related concepts. As for platform models, many works have been done on homogeneous platforms, including all kinds of task models; less have been done for heterogeneous platforms. Most existing works tend to assume a too "ideal" power consumption model. Thus, the proposed scheduling algorithms may not be suitable for practical systems. For task models, we notice that only relatively simple tasks are considered thoroughly on heterogeneous multiprocessor platforms, namely, frame-based tasks and periodic tasks. Tasks with precedence constraints and sporadic tasks also need more research effort. For other related concepts, we have mentioned that they can have great influences on both schedulability and energy consumption. These concepts reflect how practical platforms and tasks may have additional constraints and requirements for energy-aware scheduling algorithms.

References

1. S.M. Martin, K. Flautner, T. Mudge, D. Blaauw, Combined dynamic voltagescaling and adaptive body biasing for lower power microprocessors under dynamic workloads, in *Proceedings of IEEE/ACM International Conference on Computer-Aided Design*, San Jose, November 2002, pp. 721–725

2. C.-Y. Yang, J.-J. Chen, T.-W. Kuo, An approximation algorithm for energy-efficient scheduling on a chip multiprocessor, in *Proceedings of Design, Automation and Test in Europe Conference*, vol. 1, Munich, March 2005, pp. 468–473

3. J.-J. Chen, H.-R. Hsu, K.-H. Chuang, C.-L. Yang, A.-C. Pang, T.-W. Kuo, Multiprocessor energy-efficient scheduling with task migration considerations, in *Proceedings of the 16th Euromicro Conference on Real-Time Systems*, Catania, June–July 2004, pp. 101–108

4. J.-J. Chen, T.-W. Kuo, Multiprocessor energy-efficient scheduling for real-time tasks with different power characteristics, in *Proceedings of International Conference on Parallel Processing*, Oslo, June 2005, pp. 13–20

5. F. Kong, W. Yi, Q. Deng, Energy-efficient scheduling of real-time tasks on cluster-based multicores, in *Proceedings of Design, Automation Test in Europe Conference and Exhibition*, Grenoble, March 2011, pp. 1–6

6. K. Li, Performance analysis of power-aware task scheduling algorithms on multiprocessor computers with dynamic voltage and speed. IEEE Trans. Parallel Distrib. Syst.**19**(11), 1484–1497:2008

7. K. Li, Design and analysis of heuristic algorithms for power-aware scheduling of precedence constrained tasks, in *Proceedings of IEEE International Symposium on Parallel and Distributed Processing Workshops and Phd Forum*, Anchorage, May 2011, pp. 804–813

8. P. de Langen, B. Juurlink, Leakage-aware multiprocessor scheduling for low power, in *Proceedings of the 20th International Parallel and Distributed Processing Symposium*, Rhodes, April 2006, 8pp

9. P. de Langen, B. Juurlink, Trade-offs between voltage scaling and processor shutdown for low-energy embedded multiprocessors, in *Proceedings of the 7th International Conference on Embedded Computer Systems: Architectures, Modeling, and Simulation*, Samos, 2007, pp. 75–85

10. H. Aydin, Q. Yang, Energy-aware partitioning for multiprocessor real-time systems, in *Proceedings of International Parallel and Distributed Processing Symposium*, Nice, April 2003, 9pp

11. J.-J. Chen, T.-W. Kuo, Energy-efficient scheduling of periodic real-time tasks over homogeneous multiprocessors, in *The Second International Workshop On Power-Aware Real-Time Computing*, Jersey City, 2005, pp. 30–35

D. Li and J. Wu, *Energy-aware Scheduling on Multiprocessor Platforms*,
SpringerBriefs in Computer Science, DOI 10.1007/978-1-4614-5224-9,
© The Author(s) 2013

12. J.-J. Chen, H.-R. Hsu, T.-W. Kuo, Leakage-aware energy-efficient scheduling of real-time tasks in multiprocessor systems, in *Proceedings of the 12th IEEE Real-Time and Embedded Technology and Applications Symposium*, San Jose, April 2006, pp. 408–417

13. C. Xian, Y.-H. Lu, Z. Li, Energy-aware scheduling for real-time multiprocessor systems with uncertain task execution time, in *Proceedings of the 44th ACM/IEEE Design Automation Conference*, San Diego, June 2007, pp. 664–669

14. W.Y. Lee, Energy-saving dvfs scheduling of multiple periodic real-time tasks on multi-core processors, in *Proceedings of the 13th IEEE/ACM International Symposium on Distributed Simulation and Real Time Applications*, Singapore, 2009, pp. 216–223

15. V. Devadas, H. Aydin, Coordinated power management of periodic real-time tasks on chip multiprocessors, in *Proceedings of International Green Computing Conference*, Chicago, August 2010, pp. 61–72

16. E. Seo, J. Jeong, S. Park, J. Lee, Energy efficient scheduling of real-time tasks on multicore processors. IEEE Trans. Parallel Distrib. Syst. **19**(11), 1540–1552, 2008

17. D.-S. Zhang, F.-Y. Chen, H.-H. Li, S.-Y. Jin, D.-K. Guo, An energy-efficient scheduling algorithm for sporadic real-time tasks in multiprocessor systems, in *Proceedings of IEEE 13th International Conference on High Performance Computing and Communications*, Banff, Alberta, September 2011, pp. 187–194

18. G. Zeng, T. Yokoyama, H. Tomiyama, H. Takada, Practical energy-aware scheduling for real-time multiprocessor systems, in *Proceedings of the 15th IEEE International Conference on Embedded and Real-Time Computing Systems and Applications*, Beijing, August 2009, pp. 383–392

19. W. Yuan, K. Nahrstedt, Energy-efficient soft real-time cpu scheduling for mobile multimedia systems, in *Proceedings of the 9th ACM Symposium on Operating Systems Principles*, Bolton Landing, 2003, pp. 149–163

20. H. Cho, B. Ravindran, E.D. Jensen, An optimal real-time scheduling algorithm for multiprocessors, in *Proceedings of the 27th IEEE International Real-Time Systems Symposium*, Rio de Janerio, December 2006, pp. 101–110

21. V. Nelis, J. Goossens, R. Devillers, N. Navet, Power-aware real-time scheduling upon identical multiprocessor platforms, in *Proceedings of IEEE International Conference on Sensor Networks, Ubiquitous and Trustworthy Computing*, Taichung, June 2008, pp. 209–216

22. V. Nelis, J. Goossens, Mora: an energy-aware slack reclamation scheme for scheduling sporadic real-time tasks upon multiprocessor platforms, in *Proceedings of the 15th IEEE International Conference on Embedded and Real-Time Computing Systems and Applications*, Beijing, August 2009, pp. 210–215

23. M. Bertogna, M. Cirinei, G. Lipari, Improved schedulability analysis of edf on multiprocessor platforms, in *Proceedings of the 17th Euromicro Conference on Real-Time Systems*, Palma de Mallorca, Balearic Islands, July 2005, pp. 209–218

24. J. Goossens, S. Funk, S. Baruah, Priority-driven scheduling of periodic task systems on multiprocessors, in *Proceedings of IEEE International Real-Time Systems Symposium*, Cancun, vol. 25, September 2003, pp. 187–205

25. S. Funk, Lre-tl: an optimal multiprocessor algorithm for sporadic task sets with unconstrained deadlines, in *Proceedings of IEEE International Real-Time Systems Symposium*, San Diego, vol. 46, no. 3. (Kluwer, Norwell, 2010), pp. 332–359

26. W. Sun, T. Sugawara, Heuristics and evaluations of energy-aware task mapping on heterogeneous multiprocessors, in *Proceedings of IEEE International Symposium on Parallel and Distributed Processing Workshops and Phd Forum*, Alaska, May 2011, pp. 599–607

27. D. Li, J. Wu, Energy-aware scheduling for frame-based tasks on heterogeneous multiprocessor platforms, in *Proceedings of International Conference on Parallel Processing*, Pittsburgh, September 2012

28. S. Boyd, L. Vandenberghe, *Convex Optimization* (Cambridge University Press, Cambridge, 2004)

29. Y.C. Lee, A. Y. Zomaya, Minimizing energy consumption for precedence-constrained applications using dynamic voltage scaling, in *Proceedings of the 9th IEEE/ACM International Symposium on Cluster Computing and the Grid*, Shanghai, May 2009, pp. 92–99

30. C.-M. Hung, J.-J. Chen, T.-W. Kuo, Energy-efficient real-time task scheduling for a dvs system with a non-dvs processing element, in *Proceedings of the 27th IEEE International Real-Time Systems Symposium*, Rio de Janerio, December 2006, pp. 303–312

31. C.-Y. Yang, J.-J. Chen, T.-W. Kuo, L. Thiele, An approximation scheme for energy-efficient scheduling of real-time tasks in heterogeneous multiprocessor systems, in *Proceedings of Design, Automation Test in Europe Conference and Exhibition*, Nice, April 2009, pp. 694–699

32. J.-J. Chen, L. Thiele, Task partitioning and platform synthesis for energy efficiency, in *Proceedings of the 15th IEEE International Conference on Embedded and Real-Time Computing Systems and Applications*, Beijing, 2009, pp. 393–402

33. J.-J. Chen, T.-W. Kuo, Allocation cost minimization for periodic hard real-time tasks in energy-constrained dvs systems, in *Proceedings of the 2006 IEEE/ACM International Conference on Computer-Aided Design*, San Jose, 2006, pp. 255–260

34. S.K. Baruah, Partitioning real-time tasks among heterogeneous multiprocessors, in *Proceedings of International Conference on Parallel Processing*, vol. 1, Montreal, August 2004, pp. 467–474

35. J.-J. Chen, C.-F. Kuo, Energy-efficient scheduling for real-time systems on dynamic voltage scaling (dvs) platforms, in *Proceedings of the 13th IEEE International Conference on Embedded and Real-Time Computing Systems and Applications*, Daegu, August 2007, pp. 28–38

36. J.-J. Chen, C.-Y. Yang, T.-W. Kuo, C.-S. Shih, Energy-efficient real-time task scheduling in multiprocessor dvs systems, in *Proceedings of Asia and South Pacific Design Automation Conference*, Pacifico Yokohama, January 2007, pp. 342–349

37. C.-Y. Yang, J.-J. Chen, T.-W. Kuo, L. Thiele, Energy reduction techniques for systems with non-dvs components, in *Proceedings of IEEE Conference on Emerging Technologies Factory Automation*, Palma de Mallorca, September 2009, pp. 1–8

38. H. El Ghor, M. Chetto, R. Hage Chehade, G. Nachouki, Towards multiprocessor scheduling in distributed real-time systems under energy constraints, in *Proceedings of International Conference on Advances in Computational Tools for Engineering Applications*, Beirut, July 2009, pp. 355–361

39. B.A. Allavena, A. Allavena, D. Moss, Scheduling of frame-based embedded systems with rechargeable, in *Workshop on Power Management for Real-Time and Embedded Systems (in conjunction with RTAS)*, Taipei, 2001

40. R.I. Davis, A. Burns, A survey of hard real-time scheduling for multiprocessor systems. J. ACM Comput. Surv. **43**(4), 35:1–35:44, 2011

41. C.L. Liu, J.W. Layland, Scheduling algorithms for multiprogramming in a hard real-time environment. J. ACM **20**, 46–61, 1973